Springer Water

The book series Springer Water comprises a broad portfolio of multi- and interdisciplinary scientific books, aiming at researchers, students, and everyone interested in water-related science. The series includes peer-reviewed monographs, edited volumes, textbooks, and conference proceedings. Its volumes combine all kinds of water-related research areas, such as: the movement, distribution and quality of freshwater; water resources; the quality and pollution of water and its influence on health; the water industry including drinking water, wastewater, and desalination services and technologies; water history; as well as water management and the governmental, political, developmental, and ethical aspects of water.

Elena Neverova-Dziopak · Zbigniew Kowalewski

Eutrophication: A Global Environmental Problem

Process Management Strategies

Springer

Elena Neverova-Dziopak
Faculty of Geo-Data Science
Geodesy, and Environmental Engineering
Department of Environmental Management
and Protection
AGH University of Krakow
Kraków, Poland

Zbigniew Kowalewski
Faculty of Geo-Data Science
Geodesy, and Environmental Engineering
Department of Environmental Management
and Protection
AGH University of Krakow
Kraków, Poland

ISSN 2364-6934 ISSN 2364-8198 (electronic)
Springer Water
ISBN 978-3-031-83925-2 ISBN 978-3-031-83926-9 (eBook)
https://doi.org/10.1007/978-3-031-83926-9

This Springer imprint is published by the registered company Springer Nature Switzerland AG
The registered company address is: Gewerbestrasse 11, 6330 Cham, Switzerland

Summary

Currently eutrophication has evolved from a local and regional manifestations to a global scale problem with very serious negative environmental, economic, social and health security consequences. Eutrophication of water bodies started attracting more and more attention from researchers since the 1970s. Despite this, it has not yet been possible to control this process in order to prevent its negative consequences. Due to the intensive economic development, population growth and urbanization, the eutrophication intensification and expansion of its geography is observed, especially in the context of changing climatic conditions. The solution to this problem is of great practical importance for the protection and exploitation of the planet's water resources, which is an integral condition for sustainable development.

The book offers the results of critical analysis of the reasons for the lack of control over this process and tangible success in preventing progressing eutrophication and its negative effects. Particular attention is paid to the critical analysis of the current state of knowledge on the mechanisms and factors of eutrophication, the existing approaches of this process managing, the legislative aspects, the methods of nutrients loads reduction, appropriate selection of wastewater treatment technologies, the issues of trophic status assessment and monitoring etc.

The authors of monograph conducted the deep holistic analysis of failures and unsatisfactory efficiency of eutrophication mitigation measures, considering the unaccounted or ignored factors of process management. On the base of such analysis the authors proposed an algorithm for the concept of integrated eutrophication management, which is necessary to achieve the desired results and undoubtedly promising.

The target auditorium of this book—a wide range of specialists in water resources management and protection, water-and-wastewater technologies, spatial planning, as well as lawyers and economists for environmental protection, medical workers, students and doctoral students.

Introduction

Always select the scale one size larger than your problem, because it is half driven from the larger side. That's the first principle of the systems approach.

—H. T. Odum

The process of eutrophication became an object of interest for limnologists back in the 1930s, and in the 1970s its consequences began to be considered a danger for water ecosystems, and methods of protection against eutrophication began to develop. As a result numerous publications appeared concerning the results on understanding and learning about this process, its mechanisms, cause-and-effect relationships, negative consequences and methods of prevention. Despite this, over the years, eutrophication has evolved from a local and regional problem to a global problem with very serious negative environmental, economic, social and health consequences.

As could be noticed, the vast majority of publications concerning eutrophication present the results of studies of individual aspects, causes or consequences of eutrophication, as well as different technical, technological and biotechnological methods of surface water protection, reclamation and other methods of prevention and mitigation. Unfortunately, we rarely find in publications the deep comprehensive systematic analysis of the reasons for the low effectiveness of various methods used to prevent and combat eutrophication, or it is very superficial. As a result, at the moment there are no holistic concepts for systemic management of eutrophication process which would be based on in-depth analysis of process management efficiency.

In this monograph, the authors make an attempt to conduct a deep holistic analysis of failures and unsatisfactory efficiency of eutrophication mitigation measures, considering the unaccounted or ignored factors of process management. On the base of such analysis the authors proposed an algorithm for integrated eutrophication management concept, which is necessary to achieve the desired results and undoubtedly promising.

Contents

List of Figures

List of Tables

Chapter 1
Techno-ecosystems—New Reality and Challenges

Abstract This chapter presents the role and importance of aquatic ecosystems for the functioning of the biosphere and as a strategic factor of human existence and civilization development. Aquatic ecosystems are presented as techno-ecosystems, the number and scale of which are increasingly expanding and acquiring the features of a global system, which includes various technical objects that interact with both nature, biosphere, and society. Effective management of these systems is the basis for harmonizing human relations with the planet's biosphere. This chapter presents the major challenges in water resources management: increased water demand, deterioration of underground and surface water quality; disproportionality of water consumption in closely interconnected strategic sectors of the economy—water-food-energy; ineffective management of water and hydraulic infrastructure; water conflicts and climate change impact.

Keywords Techno-ecosystems · Natural factors · Anthropogenic impact · Water scarcity · Water pollution · Climate change · Water management

Water resources protection is one of the most urgent and critical issues we face today. Water is an essential resource that supports life on Earth, and we cannot afford to neglect its conservation. Intensive use of water and its pollution lead to the reduction of the limited resources of clean fresh water available on the globe (only 3% of world reserves), deterioration in water quality and an increase of water deficiency. With the current scale of anthropogenic influence on the biosphere, the quality of natural water resources is formed not only as a result of natural ecosystems functioning, but also due to human activities. Moreover, this influence on the hydrosphere is very multifaceted and significant and therefore is increasingly negative for nature.

Over the past decades, the humanity has faced a number of challenges that threaten our water resources. One of them is the pollution of water bodies with domestic and industrial wastewaters, run-off from agricultural lands, as well as increased exploitation of water resources for different purposes. This brings to deterioration of water quality, disturbance of ecological balance of water ecosystems and threat to

E. Neverova-Dziopak and Z. Kowalewski, *Eutrophication: A Global Environmental Problem*, Springer Water, https://doi.org/10.1007/978-3-031-83926-9_1

human health. Climate change is also affecting water availability, leading to an even greater scale of water crisis.

Climate change as well may prove to be one of the most serious challenges. Nature plays a crucial role in mitigating its effects and adapting to new conditions. Forests, oceans, and soil are natural carbon stores that absorb significant amounts of carbon from the atmosphere. Ecosystems rich in biodiversity can adapt to changing conditions and ensure environmental sustainability. By protecting nature, we are fighting for our future, creating more sustainable and favourable conditions for ourselves and future generations. Due to the large mass of water on the Earth's surface (75%) and the peculiarities of its thermal properties, the Earth's hydrosphere regulates thermal processes, absorbing on average 77% of solar energy arriving at the Earth's surface, then transferring it into the atmosphere as a result of evaporation and subsequent condensation of water vapour (84% of the total radiation balance of the Earth), as well as by turbulent heat exchange [1]. The hydrosphere thus acts as a powerful heater of the atmosphere, shaping the climate conditions of the entire planet. By maintaining balance in the hydrosphere and preserving its biospheric functions, it is possible to maintain relative stability of the climate.

With the development of active human activity, the biosphere no longer includes only natural self-organizing systems, but also natural-anthropogenic formations. It does not fully possess the property of self-sustaining, internal stability, and therefore require human support to maintain their structure [2]. Eugene Odum, considered the "father of modern ecology", uses the term "techno-ecosystem", and he draws attention to the fundamental difference between natural systems and techno-ecosystems [3, 4].

The life of the significant part of humanity is extremely strongly connected with these "composite" systems, the number and scale of which are increasingly expanding and acquiring the features of a global system, which includes various technical objects that interact with both nature, biosphere, and society. This interaction process is very important both for humans and for natural environment (Fig. 1.1).

At the end of the nineteenth century attention was drawn to the thermal impact of power plants on water bodies and their ecosystems functioning. The influence of various industrial discharges on aquatic ecosystems also became obvious, and the development of environmental impacts assessing began. This period of a purely "environmental" approach lasted quite a long time. Thus, the focus was on the natural environment and natural ecosystem, which required protection against the influence of technical systems. This approach is, of course, extremely necessary, but already in the 1960th research related to biological interference began to develop, which was, in turn, initiated by the expansion of hydropower facilities construction in continental reservoirs and significant economic losses from the development of biofouling. This led to an understanding of the complexity of the relationships between technical and biotic systems. The focus has shifted to the problem of close interactions between biotic and technical elements in techno-ecosystems [5].

It must be especially emphasized that no technical system operates in a completely closed mode relative to the natural environment. In this system itself, only the ratios of technical and natural elements can be different. The combination of anthropogenic

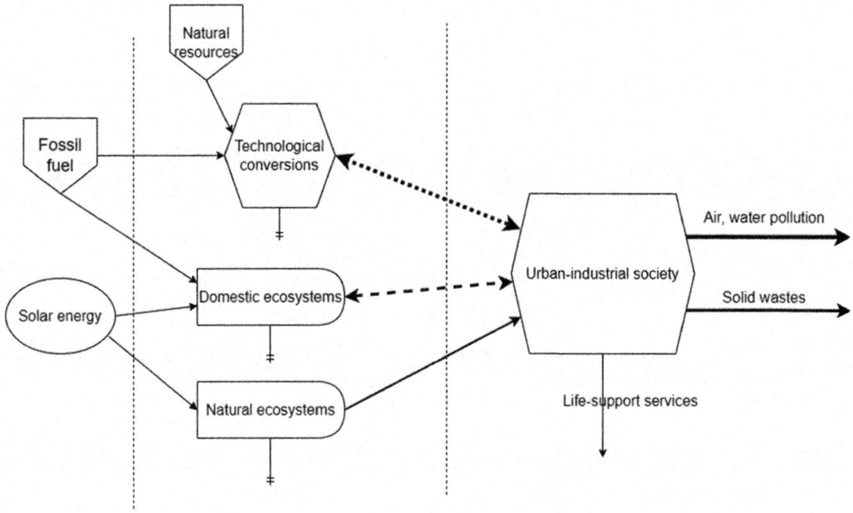

Fig. 1.1 Human-dominated techno-ecosystem and natural system (after Odum and Odum [4])

and natural elements creates additional diversity. It is obvious that techno-ecosystem, which consists of such different subsystems, will have both "mixed" properties, which will depend on the ratio of these two components and synergy of their own, unique features. Unfortunately, it is impossible to maintain the integrity and pristine state of water resources on a large scale at the current pace of economic development and scientific and technological progress. However, for the normal functioning of aquatic ecosystems and providing ourselves with water of safe quality, we must join the biotic cycle of the hydrosphere and to be not only the consumers, but also take care of the restoration and inexhaustibility of water resources and maintaining their proper quality.

The processes occurring in techno-ecosystems are dual in nature and are determined by both natural and human-caused factors. The influence of natural factors must be considered, and technical factors can be regulated. Having examined in detail the phenomena associated with over time changes in artificial reservoirs, it was pointed out significant differences in this regard between natural and techno-dependent ecosystems. If natural systems are characterized by progressive changes leading to a climax state characteristic of a given landscape-climatic zone, then in water techno-ecosystems the changes occur in several phases, each of which is associated with the characteristics of the period of operation: the phase of preparation for operation, a number of phases of relative stability and the post-technogenic phase [6]. One way or another, technical factors constantly change the course of the succession process to a greater or lesser extent. However, it should be noted, that in the absence of significant fluctuations in conditions and the level of influence of technical factors, a state of "technogenic pseudo-climax" can be established.

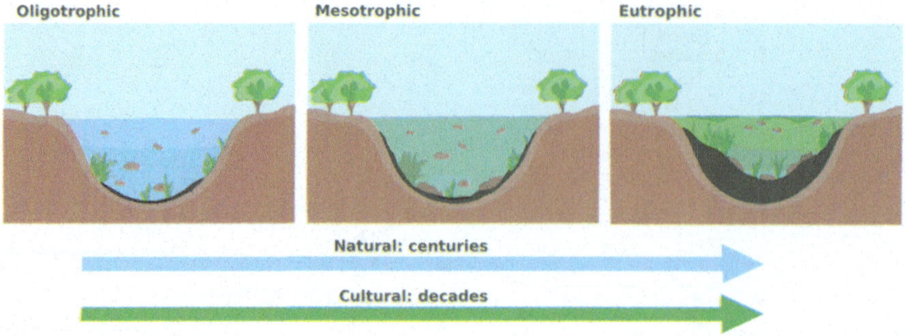

Fig. 1.2 The transformation of water ecosystem under anthropogenic impact (Torch Conservation Center https://conservetorch.org/nutrients/)

These interactions are diverse, but the influence of a techno-ecosystem on surrounding ecosystems often has a negative direction (Fig. 1.2).

Currently, it is becoming increasingly obvious that the development of anthropogenic systems connected with water bodies is impossible without considering many biotic and environmental factors. Although the important but one-sided environmental approach still prevails. The main task is not only to control, but also to develop principles and methods for managing an entire techno-ecosystem. Effective management of these systems is the basis for harmonizing human relations with the planet's biosphere. The creation of a biotic cycle in hydrosphere with the harmonious inclusion of humans in it is a long, complex process and has not yet been sufficiently scientifically developed. This requires multilateral and efficient activities based on a complex of coordinated environmental, organizational, economic, legal and technical aspects to protect water bodies from depletion, pollution and eutrophication.

1.1 Water Crisis

Water is the most important and strategic natural resource the value of which is constantly growing. Its quality and availability depend on many global (e.g. climate change) and national factors (e.g. poor water resource management).

Among the global challenges the following factors can be mentioned:

- The growth of the Earth's population, which gives rise to problems of realizing the universal right of every person to water (Fig. 1.3).
- Deterioration of underground and surface water quality, often catastrophic in some regions. The issues of transboundary water pollution are acute (Fig. 1.4).
- Energy-intensive production processes lead to an increase in electricity production which requires great amount of water. Often, its irrational, excessive use for irrigation leads to the degradation of agricultural lands, and the removal of nitrogen

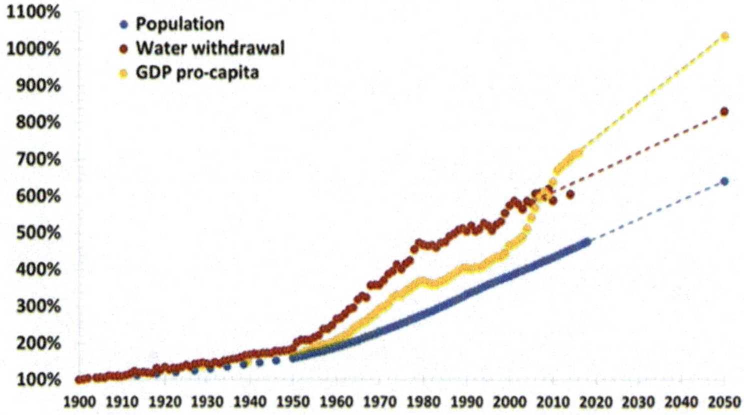

Fig. 1.3 The rate of growth in freshwater withdrawal and consumption has been even more rapid than global population growth (modified from Boretti and Rosa [7] CC BY)

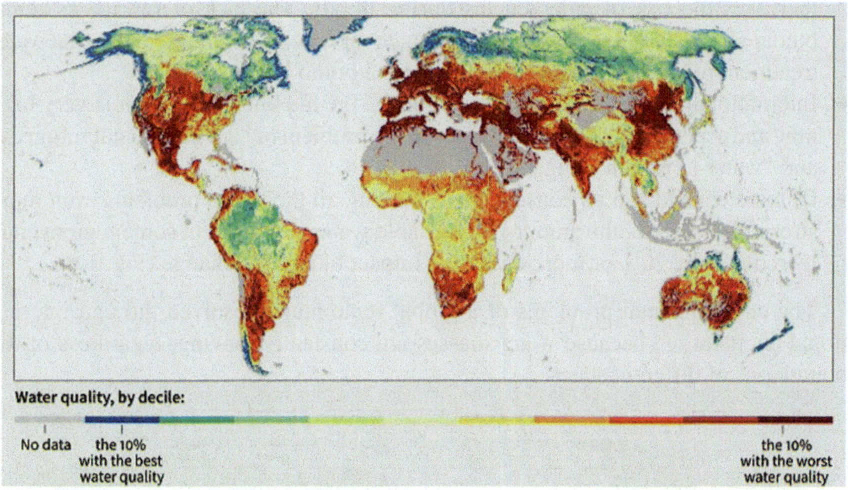

Fig. 1.4 Global water pollution map (2000–2010) (World Bank [8] CC BY)

and phosphate fertilizers from fields, creates a threat of surface water eutrophication and a decrease in fishing productivity. The sectoral approach leads to a disproportionality of closely interconnected strategic sectors of the economy—water-food-energy nexus (Fig. 1.5) and to the fact that the "battle for water" is won by energy and industry, while domestic and municipal services, agriculture and the "ecosystem services" sector lose.

- Ineffective management of water and hydraulic infrastructure is another important problem. Due to insufficient volumes of repair and restoration work, the condition of hydraulic structures deteriorates and reservoirs become silted, which, in turn,

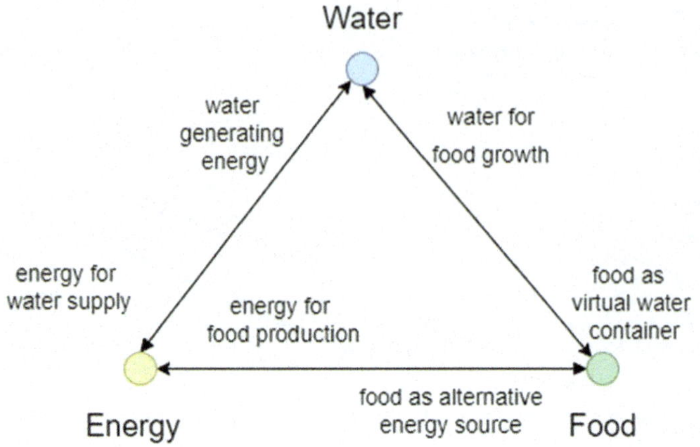

Fig. 1.5 The water-food-energy nexus food security (modified from Klingbeil and Byiringiro [9])

increases the risk of emergencies during floods. The lack of sewage systems, outdated sewer networks and inadequate operation of municipal wastewater treatment plants can be the "ecological time bomb" (Fig. 1.6).

- International conflicts over water resources, the likelihood of which is very high now and especially in the future, leads to the problem of "environmental refugees" and "water terrorism" (Fig. 1.7).
- Climate change and its consequences making all the above problems even more stressful. By 2050, the amount of available water resources in some regions may be reduced by 20% or more under the impact of climate change (Fig. 1.8).

The environmental problems of a global scale must be solved through international cooperation, because water masses are constantly moving, regardless of the boundaries of different states.

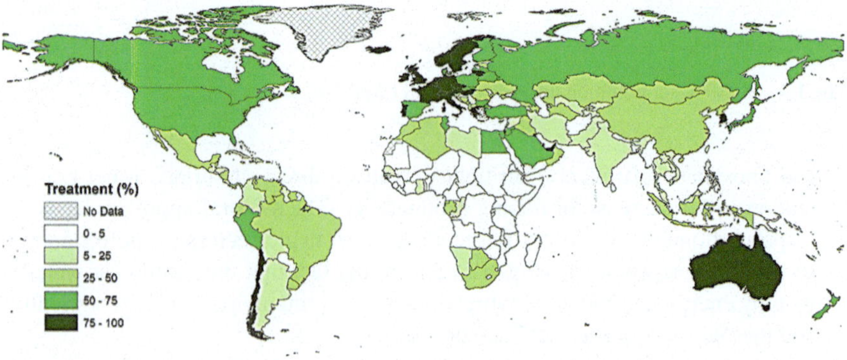

Fig. 1.6 Share of sewage undergoing treatment (World Bank [8] CC BY)

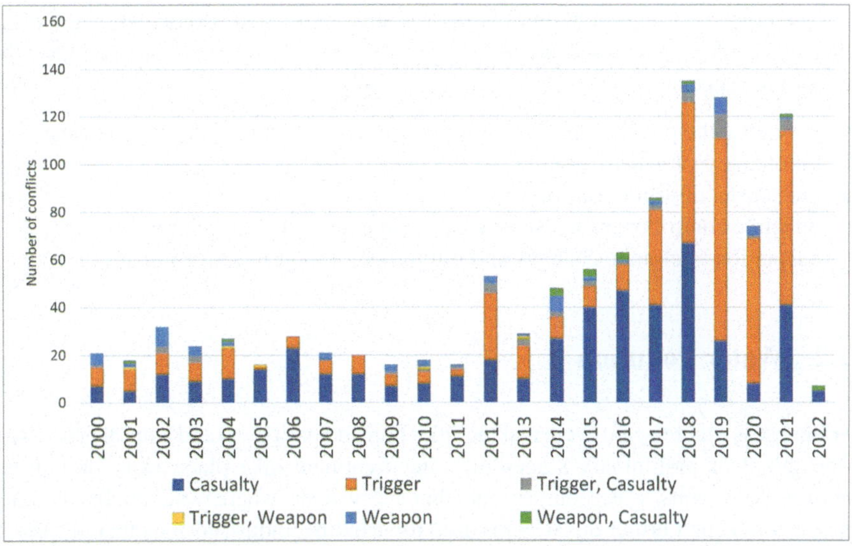

Fig. 1.7 Type of water conflicts (2000–2021) [10]

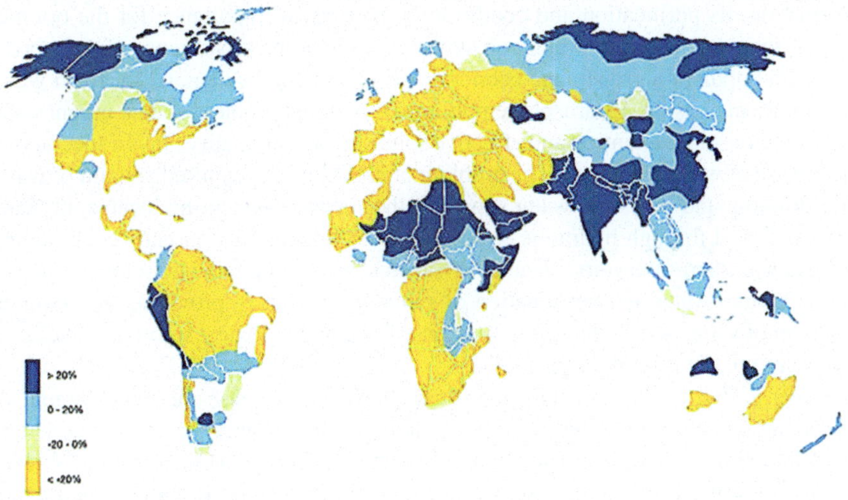

Fig. 1.8 The contribution of climate change to declining water availability (from https://www.grida.no/resources/5639 [11])

Water resource protection is not only a climatic, environmental and ethical imperative, but it also has significant economic benefits. Protecting natural resources is also important for agriculture and fisheries, which are the main sources of food for many countries. In addition, water resource conservation reduces the risks of natural disasters such as floods and droughts, which can lead to huge economic losses.

The urgency of the problem of water resource protection is dictated by the ever-increasing environmental load on all types of water resources and includes the following aspects:

- providing the population with high-quality water in the required quantities;
- rational use of water resources;
- prevention of water pollution;
- effective control over the use of water resources and their quality;
- adaptation to climate changes and mitigation of their negative impact.

1.2 Water Pollution

Water use is a large-scale cycle that includes both technological and natural processes. The initial or preliminary stages of water treatment take place in natural water bodies—reservoirs, watercourses and other ecosystems, where water quality shaping processes occur. Further this water is used for drinking, industrial and other purposes. The final stages of water purification also occur in nature, i.e. in those water bodies where the used water is discharged after it has undergone various treatment processes in special facilities. Therefore, understanding the natural mechanisms of water quality formation, its purification and conditioning is of great importance for the optimal organization of the entire water consumption cycle in the broad sense of the word.

With population growth and economic development, the demand for fresh water is rapidly increasing. Hydrosphere pollution is a global problem and an urgent issue that requires close attention and effective solutions. Water quality plays a special role in the use of water resources. The problem of providing the required amount of water for drinking and industrial water supply at the current engineering level is, in many cases, solved through hydraulic structures, river flow regulation and accumulation of water in dam reservoirs, river flow transfer from some regions to others, use of underground water and desalination of salt waters. That is, providing the required amount of water now is the result of technical and technological solutions. The issue of water quality is on the agenda.

This is due to the fact that poor-quality water can cause the development of unfavourable changes in water bodies, leading to the occurrence of infectious and non-infectious human diseases. According to the World Health Organization (WHO), about 80% of all infectious diseases in the world is associated with unsatisfactory quality of drinking water and violations of sanitary and hygienic water supply standards. The number of people with chronic diseases due to the use of contaminated water is approaching 2 billion. About 5 million people die from it every year. Polluted water brings to the death of more people than disasters and conflicts (Fig. 1.9).

All negative processes taking place in the river basins affect the state of natural waters. In this sense, natural waters can be considered simultaneously both as an integrator and as an indicator of the processes occurring in a given territory. It is associated not only with toxic waste emissions from industrial enterprises and other human activities, but also with natural phenomena. Natural sources of water pollution

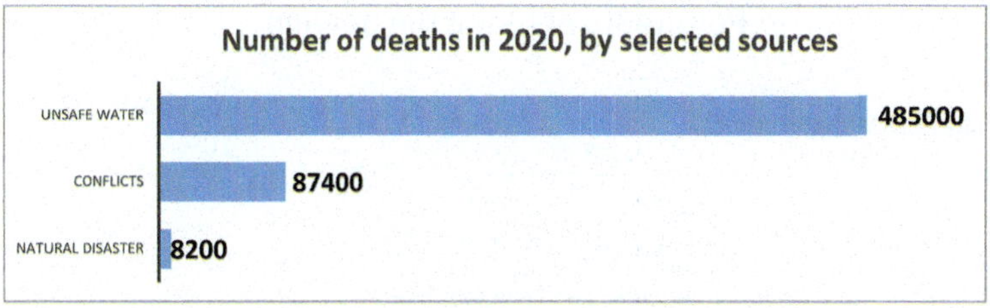

Fig. 1.9 The number of water related deaths [12]

are: leaching, water erosion of coastal soil; decay of the remains of dead animals or plants; active volcanic activity; release of metabolic products of organisms. The natural and anthropogenic pollution factors are shown at Fig. 1.10

The final link in the technogenic migration of chemical elements is the World Ocean. Its progressive pollution is largely irreversible and can lead to global consequences. The rating of the most dangerous anthropogenic pollution factors are: municipal and industrial wastewaters, atmospheric precipitation, agriculture run-off, oil leaks, thermal pollution and solid wastes (Fig. 1.11).

The problem of water pollution has been pressing for a long time and many states are addressing it in different ways and on different scales. At the end of the last century, it became obvious, that the deterioration of the quality of surface waters and the loss of their resource value are associated not only with direct pollution of water bodies coming from various sources, but also with a disruption of the ecological balance, leading to the breakdown of self-purification mechanisms and the

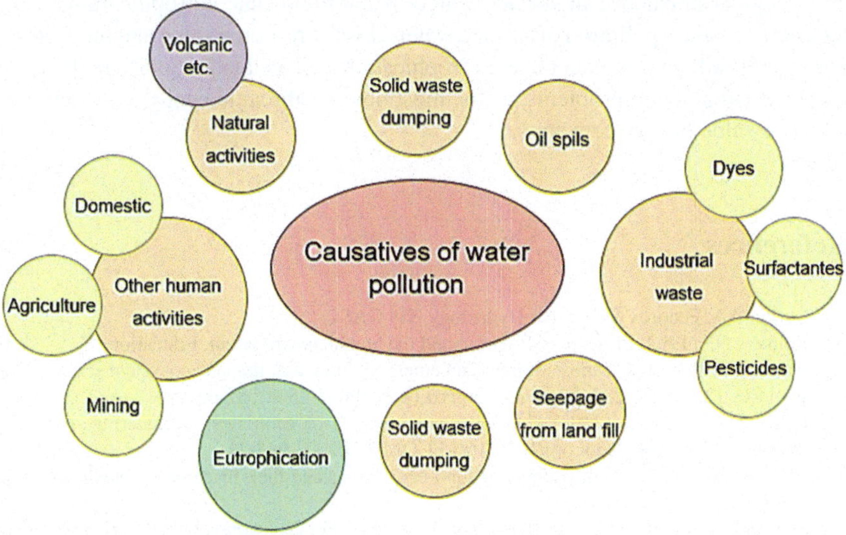

Fig. 1.10 The natural and anthropogenic pollution factors (modified form Zahid et al. [13])

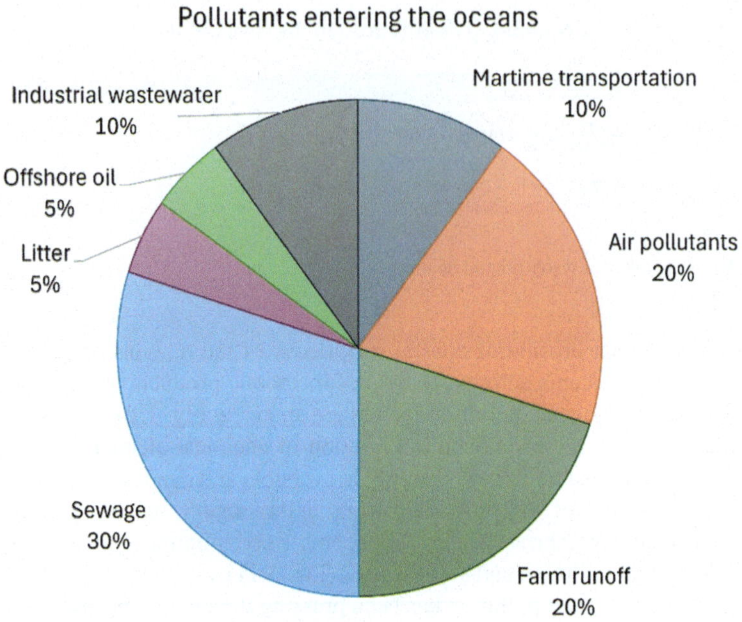

Fig. 1.11 Pollutants entering the oceans [14]

emergence of secondary negative effects. One of the main reasons for the degradation of water bodies is insufficient knowledge of human activities impact on aquatic ecosystems functions and intra-reservoir processes. The most obvious manifestation of ecological imbalance in surface waters is anthropogenic eutrophication, which leads to secondary pollution of surface waters, loss of resource value and interference with almost all types of water use. Eutrophication is placed in first place in the list of global environmental problems, as the most obvious threat to ecological safety and resource value of surface waters.

References

1. Savkin AA, Fedorov SV (2010) Hydrology. SPbGASU
2. Reimers NF (1992) Protection of nature and the human environment. Education
3. Naveh Z (1982) Landscape ecology as an emerging branch of human ecosystem science. Adv Ecol Res 12(C):189–237. https://doi.org/10.1016/S0065-2504(08)60079-3
4. Odum HT, Odum B (2003) Concepts and methods of ecological engineering. Ecol Eng 20(5):339–361. https://doi.org/10.1016/J.ECOLENG.2003.08.008
5. Protasov AA (1997) About aquatic techno-ecosystems and their place in the biosphere. J Sib Fed Univ
6. Suzdaleva A (2020) Ecological globalistics and the paradigm of world civilization development. E3S Web Conf 217:11003. https://doi.org/10.1051/e3sconf/202021711003

7. Boretti A, Rosa L (2019) Reassessing the projections of the world water development report. Npj Clean Water 2(1)
8. World Bank (2010) World development report 2010
9. Klingbeil R, Byiringiro F (2013) Food security, water security, improved food value chains for a more sustainable socio-economic development. https://doi.org/10.13140/2.1.2616.1443
10. Gleick P (2022) Water conflicts continue to worsen worldwide. Pacific Institute
11. Rekacewicz P (2006) The contribution of climate change to declining water availability. GRID Arendal
12. Herre B, Rodes-Guirao L (n.d.) War and peace. Published Online at OurWorldInData.Org. https://Ourworldindata.Org/War-and-Peace
13. Zahid M, Nadeem N, Hanif MA, Bhatti IA, Bhatti HN, Mustafa G (2019) Metal ferrites and their graphene-based nanocomposites: synthesis, characterization, and applications in wastewater treatment, pp 181–212. https://doi.org/10.1007/978-3-030-16439-3_10
14. Melnyk O, Onishchenko S, Onishchenko O (2023) Development measures to enhance the ecological safety of ships and reduce operational pollution to the environment. Sci J Silesian Univ Technol Ser Transp 118:195–206. https://doi.org/10.20858/sjsutst.2023.118.13

Chapter 2
Historical Aspects

Abstract Records of fossil akinetes of Aphanizomenon and Anabaena (Cyanobacteria) in the laminated sediments of Lake Gosciaz are interpreted. Increasing human impact in the catchment area of the lake (as can be interpreted from the pollen records of human impact indicators) apparently had its effect on the trophic conditions of the lake water: from ca. 1000 AD on fosil akinetes of Aphinizomenon and Anabaena are present in enormous quantitites in the sediment. The increases of the Cyanobacteria are interpreted as the effect of an intensiication of farming and land fertilization in the area around Lake Gosciaz, causing eutrophication of the lake. Phosphorus enrichment form effluent and excreta in the catchment area of the lake at times will have become so high that N-limited growth conditions occurred. In such conditions Cynoobacteria capable of nitrogen fixation (namely Alphanizomenon and Anabaena) could bloom. The deposits of the last ca. two centuries are characterised by the successive appearance of Pediatrum boryanum, Tetraedron minimum, Coelastrum cf. reticulatum, Botryococcus, Scenedesmus, Spirogyra, Gloeotrichia and Staurastrum manfeldtii, whereas Cyanobacteria show a decline. This phytoplankton succession could be interpreted in terms of competition for nutrients and light: higher eutrophication levels and higher turbidity caused a decline of available light, and as a consequence less enegry was avilable for the energyu consuming process o nitrogen fixation by Aphanizomenon and Anabaena.

Keywords Ancient times · Inhabitant areas · Paleolimnological studies · Bottom sediments · First Egyptians plague

Richard A. Vollenweider wrote, that eutrophication is the oldest water quality problem caused by anthropogenic activities [1]. This fact has been confirmed by numerous paleolimnological studies of bottom sediments. People have always settled near a sources of water—rivers, lakes and sea coasts. Pollutants from inhabited areas found their way into waters, contributing to the increased development of aquatic vegetation. The results of the investigations of different researches on lake sediments, generalized by G. E. Hutchinson allowed to conclude, that the eutrophication phenomenon had already occurred in very ancient times: in many small lakes a

E. Neverova-Dziopak and Z. Kowalewski, *Eutrophication: A Global Environmental Problem*, Springer Water, https://doi.org/10.1007/978-3-031-83926-9_2

Fig. 2.1 Lago di Monterosi (Stifone, CC BY-SA 4.0 via Wikimedia Commons)

succession from more inorganic sediment, containing oligotrophic-indicator fossils, to a more organic sediment, containing eutrophic-indicator fossils, is usual [2]. Also the study of changes in diatoms in Linsley Pond in Connecticut "certainly indicate that the lake was beginning to respond to the results of European colonization" [3].

Very interesting example can be presented by the history of Lago di Monterosi eutrophication happened sometime between about 390 and 150 B.C.—the most probable period of the Via Cassia building by the Romans in 171 B.C. This little lake, north of Rome on the Via Cassia, occupies the explosion crater, which was formed about 26,000 years ago, was shallow and unproductive. But an event occurred which caused the increase of water level and bloom of blue-green algae (probably *Aphanizomenon*) and led to a rapid sedimentation of organic and inorganic matter [4, 5]. The authors of the paper have believed, that the cause of dramatic eutrophication recorded in the sediments was cutting of wood in the lake catchment area for road construction purposes. So, the process was induced by anthropogenic factors (Fig. 2.1).

Another interesting example was described by W. Ohle, concerning the Great Plön Lake in Germany [6]. The study of the bottom sediments indicates a slow increase in organic sedimentation from post-glacial times to the Middle Ages. But suddenly, in or about A.D. 1256, there was a tenfold increase in the intensification of inorganic and organic matter sedimentation process. The author states, that the main cause of such dramatic changes in medieval times is very probably connected with artificial flooding of land when German colonists which began to enter this

Fig. 2.2 The first plague: water is changed into blood, by nineteenth century French painter James Tissot (CC BY-SA 4.0 via Wikimedia Commons)

previously Slavonic territory. Once again, this example confirms the contribution of anthropogenic factors to "galloping eutrophication".

The occurrence of this phenomenon observed by ancient people led to the emergence of legends and myths. One of them is the first plague upon the Egyptians, when Moses struck the river Nile with his staff, turning its waters to blood (Fig. 2.2). At the same time, his brother Aaron performed an identical transformation in the canals, tributaries, ponds and reservoirs throughout Egypt. This type of phenomena still takes place on the Nile in Egypt (Fig. 2.3).

In 1957, scholar Greta Hort suggested it was due to silt from the river's southern tributaries which carried flagellate protozoa [7]. But in the 1990s, analysis by H. M. Duncan Hoyte, John S. Marr and Curtis D. Malloy suggested the miracle was due to an infestation of toxic red algae [8].

The data obtained in investigating the sedimentary record of Lake Murten, Switzerland which experienced several periods of intensive human land-use over the past 2000 years, show that "cultural eutrophication is not exclusively a modern phenomenon, but already took place 2000 years ago during the Roman period" [9].

Fig. 2.3 A satellite image of the Nil using infrared technology colours the river in red (Courtesy European Space Agency, April 2016CC BY-SA 3.0 IGO or ESA Standard Licence)

The first negative anthropogenic impacts are most often linked to deforestation of catchment area and soil erosion, followed by contamination and disturbance of aquatic ecosystems, and the most spectacular symptom of which were mass algae blooms. The evolution of anthropogenic impacts on water ecosystem is presents at Fig. 2.4.

Paleolimnological records for Holocene sediment from European lakes provide evidence of cultural eutrophication dating as far back as the Neolithic period. The data obtained for the lakes in Germany, Finland, Sweden, Estonia, Iberian Peninsula, Northern French confirm, that mainly soil erosion due to agricultural activities in catchment areas was the main propulsion for cultural eutrophication in Europe. The increased production of *Oscillatoria rubensces* in Lake Marten, known as "Blood of the Burgundies", occurred in 1820 [10].

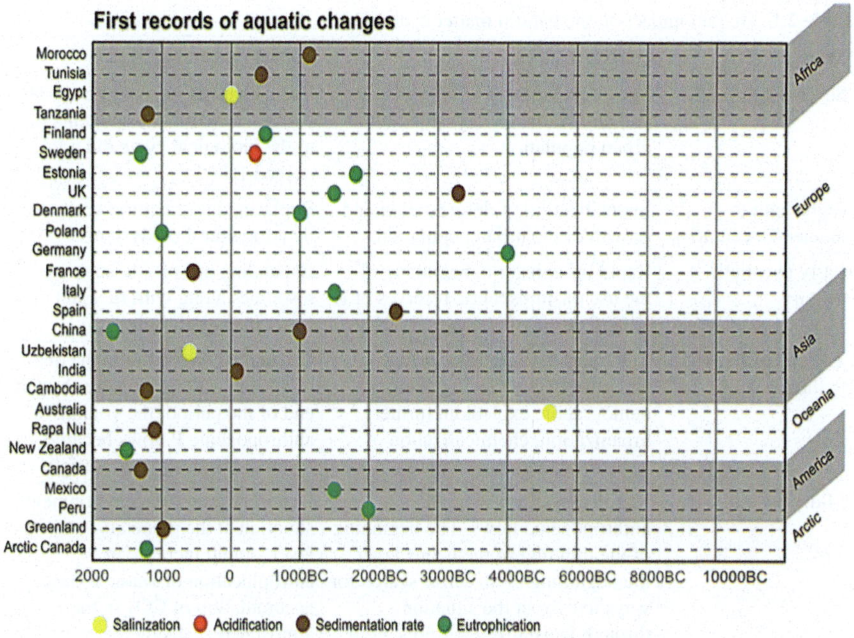

Fig. 2.4 First records of aquatic changes recorded in palaeolimnological archives (modified from Dubois et al. [10])

The results of sediment investigations from three sites along the east-coast of Sweden, north-western Baltic Proper allowed to confirm the appearance of hypoxia zone in the open Baltic Sea during the Medieval Climate as the result of eutrophication caused by anthropogenic transformation of land [11].

Increased production in Swiss lakes started at the turn of the XIX–XX century [11]. The beginning of the XX century was marked by the greatest anthropogenic pressure on the aquatic ecosystems. Eutrophication in Europe accelerated due to population increase and agricultural and industrial sectors development.

Technogenesis, which by the end of the twentieth century had reached a scale commensurable with natural geochemical processes, significantly disrupts the natural cycle of substances in the biosphere, which can fully be called the cause of globally increasing eutrophication. The main reason for changes in energy and material cycles is considered to be the growth of the planet's population and changes in technologies in agriculture and industry. Six stages of human impact on natural matter cycles can be distinguished (Table 2.1). The basis for the division into these stages was the analysis of changes occurring in water quality in inland water bodies, particularly in connection with the process of anthropogenic eutrophication, which causes an increase in the flow of nutrients into the aquatic ecosystems. The most reliable estimates are of changes in phosphorus fluxes: thus, the total planetary phosphorus load on inland waters has currently increased by 2.5 times compared to the natural one.

Table 2.1 Human impact on the natural matter cycles [12]

Stage	Human activities	Impact on natural cycles
Seventeenth century	Slow development of agriculture and animal husbandry, small total urban population	Local changes in ecosystems. The contribution of anthropogenic flows to the biological cycle did not exceed 1%
Eighteenth to nineteenth centuries	Intensification of deforestation and growth of the urban population	Replacement of natural ecosystems by agricultural ecosystems
Early twentieth century	Rapid urbanization, beginning of the use of lithospheric reserves of P, K, S, fixation of atmospheric N	Expanding use of lithosphere resources along with fuel and ores on an industrial scale
The end of the 40s to the 60s	The rapid development of farms, activities in developed countries, demographic explosion, further urbanization, chemicalization of agriculture and everyday life	Gradual deterioration in the quality of natural waters, air, soils: by the end of the period, the volume of anthropogenic P flows became equal to flows of natural origin
Mid. 60s–mid. 80s	Significant increase in all anthropogenic flows. The beginning of the thinning of the ozone layer, the beginning of an active search for ways to correct the situation through legislative measures, both in individual states and at the level of interstate projects	Deterioration of the environmental situation at the planetary level—rapid increase in eutrophication of water bodies, accumulation of CO_2 in the atmosphere, causing the greenhouse effect
Middle–late 80s	The transition to technologies with maximum isolation, attempts to develop international mechanisms for regulating economic activity	Critical situations with water and air quality in certain regions, the beginning of instrumental registration of global changes
Twenty-first century	At present, it is practically impossible to quantify the actual "natural" and "anthropogenic" cycles. There has been a significant replacement of natural ecosystems with artificially created ones	Modern planetary cycles of organic matter and nutrients are the result of long-term anthropogenic interference in the natural cycle. The completeness of the phosphorus cycle decreased from 0.98–0.99 to 0.5–0.6 for individual territories and to 0.91 in general

All changes associated with the flow of organic and biogenic substances into hydrosphere create conditions for the development of the global process of anthropogenic eutrophication. The closedness of cycles is broken—the biological cycle is transformed into a transit pattern of development: lithosphere—economic activity—hydrosphere. An increase in the productivity of water bodies is observed in almost all developed and less developed countries, leading to a radical restructuring of aquatic ecosystems.

References

1. Vollenweider RA, Rinaldi A, & Montanari G (1992) Eutrophication, structure and dynamics of a marine coastal system: results of ten-year monitoring along the Emilia-Romagna coast (Northwest 90 Adriatic Sea). In Marine Coastal Eutrophication. Elsevier. 63–106. https://doi.org/10.1016/B978-0-444-89990-3.50014-6
2. Hutchinson GE (1973) Marginalia: Eutrophication: The Scientific Background of a Contemporary Practical Problem. American Scientist, JSTOR, 61(3):269–279. http://www.jstor.org/stable/27843785. Accessed 26 Feb. 2025.
3. Patrick R (1943) The Diatoms of Linsley Pond, Connecticut. In: Proceedings of the Academy of Natural Sciences of Philadelphia, 95:53–110. JSTOR, http://www.jstor.org/stable/4064343. Accessed 26 Feb. 2025.
4. Hutchinson GE et al (1970) Ianula: An Account of the History and Development of the Lago Di Monterosi, Latium, Italy. Transactions of the American Philosophical Society, JSTOR, 60(4):1–178. https://doi.org/10.2307/1005996. Accessed 26 Feb. 2025.
5. van Geel B, Mur LR, Ralska-Jasiewiczowa M, Goslar T (1994) Fossil akinetes of Aphanizomenon and Anabaena as indicators for medieval phosphate-eutrophication of Lake Gosciaz (Central Poland), Review of Palaeobotany and Palynology, ISSN 0034-6667, 83(1-3):97-105, https://doi.org/10.1016/0034-6667(94)90061-2. https://www.sciencedirect.com/science/article/pii/0034666794900612
6. Ohle W (1972) Die Sedimente der Grossen Pl ner Sees als Dokumente der Zivilisation. Jahrbuch j. Heimatkunde 2
7. Hort G (1957) The plagues of Egypt. 69(1–4):84–103. https://doi.org/10.1515/zatw.1957.69.1-4.84
8. Marr JS, Malloy CD (1996) An epidemiologic analysis of the ten plagues of Egypt. Caduceus (Springfield, Ill) 12(1):7–24
9. Haas M, Baumann F, Castella D, Haghipour N, Reusch A, Strasser M, Eglinton TI, Dubois N (2019) Roman-driven cultural eutrophication of Lake Murten, Switzerland. Earth Planet Sci Lett 505:110–117. https://doi.org/10.1016/J.EPSL.2018.10.027
10. Dubois N, Saulnier-Talbot É, Mills K, Gell P, Battarbee R, Bennion H, Chawchai S, Dong X, Francus P, Flower R, Gomes DF, Gregory-Eaves I, Humane S, Kattel G, Jenny J, Langdon P, Massaferro J, McGowan S, Mikomägi A, Valero-Garcés B et al (2018) First human impacts and responses of aquatic systems: a review of palaeolimnological records from around the world. Anthr Rev 5(1):28–68. https://doi.org/10.1177/2053019617740365
11. Gerdeaux D, Anneville O, Hefti D (2006) Fishery changes during re-oligotrophication in 11 peri-alpine Swiss and French lakes over the past 30 years. Acta Oecologica 30(2):161–167. https://doi.org/10.1016/j.actao.2006.02.007
12. Bonaventura R, Zito F, Russo R (2023) Special issue on "Insights on ecotoxicological effects of anthropogenic contaminants in aquatic organisms". Toxics 11(4):311. https://doi.org/10.3390/toxics11040311

Chapter 3
Negative Consequences

Abstract This chapter presents the negative impacts of eutrophication which are considered in ecological, economic and social categories. The negative impact of eutrophication on basic physical, chemical and biological properties of water is also presented. Eutrophication has a huge impact on water quality, that may constitute a health hazard for the main water consumer—humans. The main health risk factors of eutrophicated waters are presented: favourable conditions for the development of pathogenic microorganisms, disease causative agents, algotoxins and the type of their impact on the humans. Drinking water supply from the sources with a high trophic status poses a high health risk, and not only due to the presence of toxic algae. Technical and technological problems arising in the event of the need to supply the population from water sources with a high trophic level are also discussed. Particular attention was also paid to various problems of removing nutrients from wastewater before its discharge into receiving water bodies. Economic aspects related to losses and costs incurred due to eutrophication are described, as well as the impact of climate change on the development of eutrophication and feedback relationships.

Keywords Water quality · Pathogenic microorganisms · Algotoxins · Health risk · Water supply · Wastewater treatment · Economic losses · Climate changes

One of the serious problems of anthropogenic eutrophication control is that at the initial stages of development the negative changes in aquatic ecosystems are hardly noticeable due to homeostasis mechanisms and almost indistinguishable from natural ones—such as seasonal or annual fluctuations of different abiotic and biotic factors. It is worth noting, that "water blooms" arising as a result of anthropogenic impact on the ecosystem represent its adaptation reaction in new conditions, which means that eutrophication can be considered to be a new stage of ecosystem development. Considering the impact on water ecosystems, the pollution and eutrophication are closely related and often mutually conditioned. In many cases, eutrophication occurs as a result of pollution, and pollution may results from eutrophication. In natural conditions, both processes usually occur together. However, in terms of the

E. Neverova-Dziopak and Z. Kowalewski, *Eutrophication: A Global Environmental Problem*, Springer Water, https://doi.org/10.1007/978-3-031-83926-9_3

Table 3.1 Negative consequences of eutrophication in water ecosystems

Ecological	Economic and social
• Water blooms and the development of toxic algae species • Deterioration of aquatic organisms habitat (lack of oxygen, sunlight, algae toxins) • Reduction of species diversity • General ecological imbalance • Loss of biospheric functions of aquatic ecosystems	• Deterioration of water properties • Increased costs of water and wastewater treatment • Reduced costs of land and real estate as a result of the loss of their aesthetic and recreational values • Economic losses in fisheries and tourism sector • Public health hazards

direction and degree of impact on aquatic ecosystem, these processes are substantially different. It should be particularly emphasized, that the most important difference lies in the fact, that contamination is caused primarily by the discharge of harmful substances from the external sources (primary pollution), that often reduces the biological productivity of waters, while eutrophication, conversely, leads to the increased productivity of water and its secondary pollution.

In the conditions of maintaining the homeostatic mechanisms of aquatic ecosystems, these processes are self-regulating and provide the appropriate level of biological productivity and water quality in particular ecosystem. The introduction of additional anthropogenic nutrient loads to natural water environment, under the violation of self-regulation mechanisms threatens the stability of aquatic ecosystem. Natural eutrophication, as a rule, does not exceeds the homeostasis limits. But if eutrophication exceeds the limits of natural self-regulation, the changes may occur in biocenosis structures, which lead to the intensive development of blue-green algae, fish kill and other negative results. In the case of natural eutrophication, these changes are usually temporary and ecosystem renews itself [1–3].

The negative impact of eutrophication can be considered in ecological, economic and social categories (Table 3.1; Figs. 3.1, 3.2, 3.3, and 3.4).

The development of aquatic vegetation is a natural phenomenon and up to a certain level plays a positive role in water self-purification processes. However, the positive role of algae depends, to a large extent, on their concentration. When algae species multiplies to high concentrations, their positive role in water self-purification changes into self-polluting factor, leading to the disturbance of ecological equilibrium of water ecosystems, to the impoverishment of their species structure, reduction of ecosystem resistance to negative impacts and deterioration of water quality. The negative impact of eutrophication on basic physical, chemical and biological properties of water is presented in Fig. 3.5 [4].

Fig. 3.1 Massive algae and macrophyte growth in Lake Annecy, France

Fig. 3.2 Water bloom in Lake Skadar, Montenegro (author: Elena Neverova-Dziopak)

3.1 Eutrophication as a Risk Factor for Human Health

The people's right to have an access to sufficient, safe, physically accessible water for personal hygienic and domestic use is recognized by UN General Assembly and the Human Rights Council as one of the basic human rights—fundamental to everyone's health and prosperity [5].

Eutrophication has a huge impact on water quality. The negative consequences of eutrophication include primarily the deterioration of the organoleptic properties of water, an increase in the values of BOD and COD indicators, changes of the acid–base

Fig. 3.3 Blooms on the Motława River, Gdańsk, Poland (author: Michał Preisner)

Fig. 3.4 Eutrophication of Dal Lake, Kashmir, India (author: Irfan Ali)

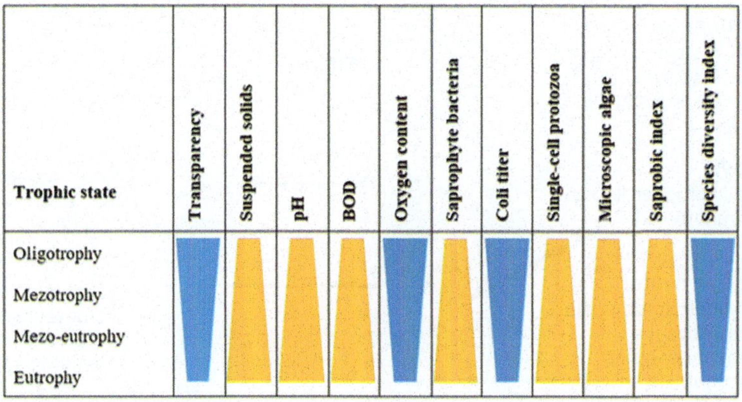

Fig. 3.5 Negative changes in water properties with trophic status increasing [4]

balance, the occurrence of oxygen deficits, the appearance of toxins, phenols and the activation of pathogenic microflora (e.g. *Poliovirus, Vibrio cholerae* and others). This is due to the fact that the above-mentioned pathogens get active in a highly alkaline environment, and just during mass blooms the pH value of water may increase to 10–12, creating conditions favorable for their development (e.g. for vibrio cholera). Research on the etiology of cholera spread allowed to record the appearance of cases of the disease during the period of mass development of cyanobacteria and in the regions of their prevalence [4].

During the mass algae bloom, some of species produce toxic metabolites that accumulate in water and demonstrate high biological activity towards various warm-blooded animals, hydrobionts and humans. Algotoxins are classified as highly toxic compounds. In acute poisoning of animals with lethal or sublethal doses, algotoxins affect the central nervous system. In chronic poisoning, the toxins disturb the metabolism of carbohydrates and proteins. The basis of the toxicodynamic effect of algotoxins is hypovitaminosis B1, hypoxemic failure and reduction of dehydrogenase activity. As a result of deep interference in metabolic processes and cellular respiration of warm-blooded animals, algotoxins have a wide spectrum of biological effects and can be classified as protoplasmic toxins with high biological activity (Fig. 3.6) [6, 7].

Many studies reported, that exposure to HABs (Harmfull Algae Blooms) could increase the risk of HAB-related diseases. The serious bloom of toxic algae was associated with the dangerous Sartlan disease (Haff illness), medical name—nutritional-toxic myoglobinuria. The most characteristic symptom of this disease is paralysis of the heart and kidney muscles of humans and domestic animals. The first case of this disease was registered in 1924–1925 among fishermen in the Frishes Gaf Bay of the Baltic Sea and described for the first time by German scientists [9]. This disease recurred three times in this region: in 1926–1927, 1932–1933 and 1939–1940. Among 1,100 cases, 18 turned out to be fatal. Further cases of this disease were registered in the former Soviet Union: in 1935–1936, 400 people became ill

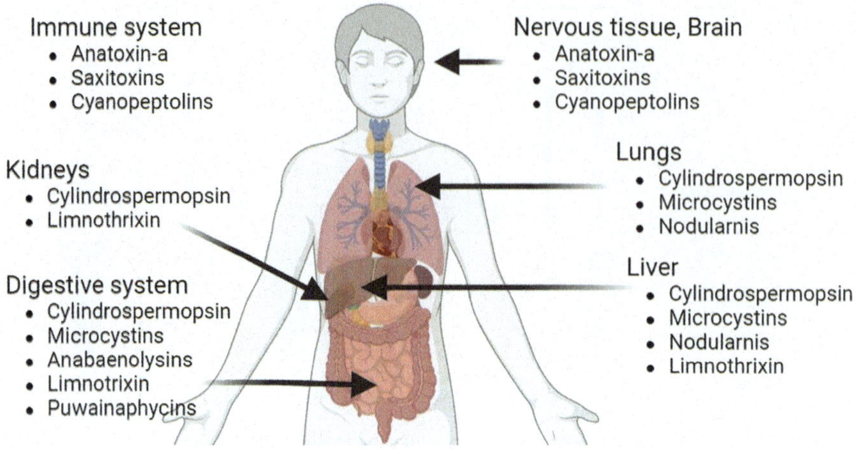

Fig. 3.6 Effects of cyanobacterial toxins on the human (after Kubickova et al. [8])

among the inhabitants of villages on Lake Yuksovsky (near St. Petersburg), and 8 died, in 1947 in the Republic of Karelia, in 1946–1947 and 1998—in towns on Lake Sartlan in the Novosibirsk Oblast, in 1960—in the Khabarovsk region, in 1970—on Lake Itkul in the Kurgan region, in 1975–1976 in the Altai region, and in 200–2001—in the Tyumen region. In all cases, the disease was caused by the toxicity of fish consumed by residents. German scientists assumed, that fish toxicity is related to the toxic bloom of cyanobacteria and the deterioration of hydrobiological conditions in lakes and dam reservoirs [10]. The cases of diseases associated with algotoxin poisoning are presented in Tables 3.2 and 3.3.

The recent research shows that eutrophication will broadly enhance infection and the pathology of human and wildlife parasites [13]. 14 states in US reported 242 harmful algal bloom (HAB) events, 63 human cases of illness, and 367 animal cases of illness for 2019. No ill persons died, 207 (56%) animal deaths were reported [14]. In South Korea the relationship between freshwater harmful algal blooms and neurodegenerative disease incidence rates were established (Fig. 3.7).

These data highlight the interconnectedness of the consequences of eutrophication process with human health, animal health and the health safety of environment. More and more often, danger warnings can be observed in water recreation areas (Fig. 3.8).

Table 3.2 Recorded cases of health problems due to poisoning with algae toxins [11]

Type	Location	Species	Symptoms	Consequences
Drinking water	USA, 1931	*Microcystis*	Gastro-enteritis	No data
	USA, 1976	*Schizotrix, Plectonema, Phormidium, Lyngbia*	Gastro-enteritis	62% of the population fed by the network became ill
	Australia, 1979	*Cylindrospermopisi*	Hepatits	141 hospitalizations
	Australia, 1981	*Microcystis*	Gastro-enteritis, liver injury	No data
	Brazil, 1988	*Anabaena, microcystis*	Gastro-enteritis	2000 people affected, 88 deaths
	Sweden, 1994	*Planktothrix agardhii*	Gastro-enteritis	121 people affected
	Brazil, 1996	*Aphanizomenon, Oscillatoria, Spirula*	Hepatits	166 people affected, 60 deaths
Recreational water	Canada, 1959	*Microcystis, Anabaena circinalis*	Gastro-enteritis, headaches, nausea, muscular pains	30 people affected
	UK, 1989	*Microcystis*	Gastro-enteritis, vomiting, sore throats	20 people affected, 2 hospitalizations

Table 3.3 Suspected and confirmed cases of human illness following exposure to cyanobacteria or algae (2007–2011) [12]

Human illness	Number of cases
Ciguatera fish poisoning	248
Rash from unknown organism or toxin	89
Illness form unknown organism or toxin	49
Microcystin poisoning	28
Other cyanobacteria or algae related illness not specified in HABISS	27
Paralytic shellfish poisoning	13
Neurotoxin shellfish poisoning	2
Anatoxin poisoning	2
Amnesic shellfish poisoning	2
Total	458

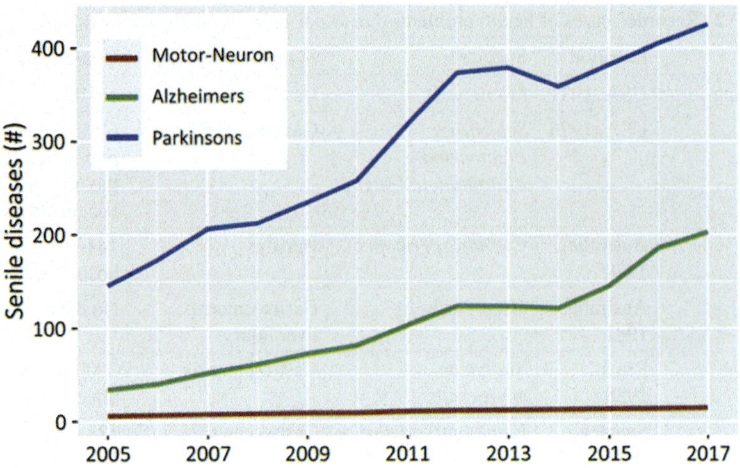

Fig. 3.7 Relationship between freshwater harmful algal blooms and neurodegenerative disease occurrence ([15] CC BY)

Fig. 3.8 Warning signs (on the base of Nevada Department of Wildlife dashboard https://www.ndow.org/blog/harmful-algae-blooms/)

3.2 Technological Problems

3.2.1 Water Supply

Drinking water supply from the sources with a high trophic status poses a high health risk, and not only due to the presence of toxic algae. When algae accumulate in water bodies under the influence of wind, the water takes on the appearance of a mass with the cream consistency. During the decomposition of algae, humic substances are formed, which contribute to the water color (green, rusty, red, blue-green and others) and which are difficult to remove in water purification plants. In dam reservoirs during water blooms, the concentration of cyanobacterial metabolites may reach the values from 1×10^{-3} to 1×10^{-8} mg/dm^3, which is sufficient to exhibit significant biological activity towards warm-blooded organisms and humans [16]. In the accumulated algae biomass, fermentation processes take place with the release of butyl spirit, acetic, lactic and butyric acid. The ammonia concentration can increase up to 100 times, COD—20–40 times, the organic nitrogen content—30–150 times, and the phenols content can exceed the permissible concentrations 10 times. Taking into account that the conditions for the development of pathogenic microflora are becoming favourable, it can be stated, that the impact of anthropogenic eutrophication on water quality is extremely great, and when the algae concentration exceeds 500 mg/l it becomes definitely negative [17].

When aquatic vegetation dies, oxygen deficits occurs leading to the fish kill. The products of primary organic matter decomposition and metabolism of various groups of bacteria (ammonia, phenols and other toxic substances) make water unsuitable for other hydrobionts. The overgrowing of the coastal zone with macrophytes, the development of periphyton and filamentous bacteria lead to disruption of the functioning of the coastal zone and important functional properties of reservoirs and watercourses.

The content of a large number of phytoplankton cells in raw water leads to a significant increase in the costs of its treatment and complications of technological systems. Water with a high content of phytoplankton cells requires the use of increased doses of coagulants. Most of the algae is removed before the filtering stage, but the risk of filter clogging remains high. Algae remaining in the water increase the content of organic substances in it, which leads to a reduction in the concentration of remaining chlorine and dissolved oxygen and favours the development of microscopic fauna [18]. Periphyton micro- and macroalgae growing on the walls of devices pose a big problem in water treatment, because special treatment of materials is required in the case, when there is no final chlorine in the water.

Not only the algae themselves, but also their metabolites cause serious complications during water treatment: they increase the demand for coagulants, favor the appearance of trihalogenomethanes (THMs), flocculation inhibitors, and dissolved gases in the water, which in turn leads to an increase in the volume of sludge in settling tanks or sudden pressure drops in the filters, the development of mold, which clogs the filters and causes its flocculation in the distribution network [19]. Algal metabolites give water an aftertaste, especially unpleasant odors. Specific odors are caused by volatile substances emitted by aquatic plants, e.g.: geosmin ($C_{12}H_{22}O$) gives the water the smell of soil and mold, 2-methylisoborneol (MIB—$C_{11}H_{20}O$)—the smell of soil and mold, hexanal ($C_6H_{12}O$) and heptanal ($C_7H_{14}O$)—a fishy smell. The dominant odor depends on the most numerous group of algae in the water (Table 3.4).

Some phenolic compounds excreted by algae react with chlorine during disinfection process and cause the chlorophenol aftertaste [18, 20].

Some products of dead organic matter decomposition, especially sulfur compounds, as well as neurotoxins (anatoxin-a, anatoxin-a(s), saxitoxin and neosaxitoxin) and hepatoxins (microcystin and nodularin) also pose a significant threat to the health of water consumers and must be removed in its treatment processes.

So, the treatment of eutrophicated water is a great technological problem, causing technical complications during water intake and purification, which are most troublesome in dam reservoirs serving as an important source of water supply to the population in many countries. Water blooms lead not only to mechanical problems at water

Table 3.4 Characterization of tastes and odors caused by algae and their metabolites [18]

Algae species	Representatives of species	Algae abundance	
		Moderate	Abundant
Cyanobacteria Cyanophyceae	*Anabena, Aphanizomenon, Gloeotrichia, Gomphosphaeria, Microcystis, Oscillatoria*	Grassy, soil-mouldy	Rots, medicines, grassy, moldy, spicy
Chlorophyceae	*Chlamidomonas, Chlorella, Cladophora, Closterium, Cosmarium, Eudorina, Pandurina, Pediastrum, Scenedesmus, Spirodyra, Staurastrum*	Grassy	Grassy, moldy, fishy
Baccillariophyceae or Diatomea	*Asterionella, Cyclotella, diatoma, Fragilaria, Melosika, Stephanodiscus, Synedra*	Grassy, grassy, spicy, geranium	Fishy, moldy
Chrysophyceae	*Dinobryon, Mallomonas, Synura*	Violets, fishy, cucumber, moldy	Fishy
Euglenophyceae	*Euglena*	–	Fishy
Dinophyceae or Dinoflagellata or Peridiniea	*Ceratium, Peridinium*	Fishy, cucumber	Rotting, medicine, fishy
Cryptophyceae	*Cryptomonas*	Violets	Violets, fishy

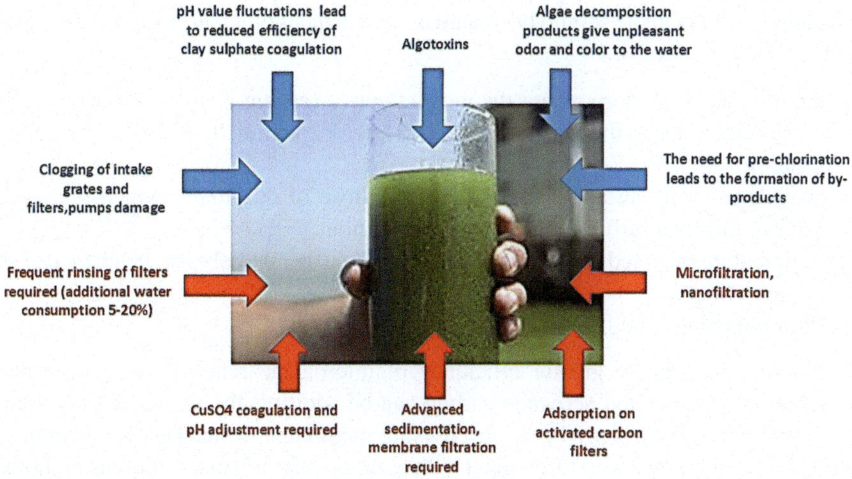

Fig. 3.9 Technical and technological problems with water supply from eutrophicated sources

intakes and purification facilities, but also deteriorate the organoleptic properties of water and their recreational and aesthetic values.

The problems of water treatment for drinking purposes and the methods of their solution are shown at Fig. 3.9.

Some authors believe that filtering, chlorination, the use of activated carbon and coagulants only slightly reduce the toxicity of algae. When coagulated, some algae form very fine flocs with poor sedimentation properties, therefore most of the phytoplankton algae are carried out onto the filters. Slow filtration is more effective in terms of removing plant biomass than fast filtration, but slow and fast filters are clogged by algae cells and the costs of their cleaning are very high. As the filtration speed increases, the efficiency of retaining algae cells decreases. When increasing the speed from 3 to 10 m/h efficiency is reduced twice. The number of residual cells also depends on the filtration time: after 40 min—75% of cells remains, after 6–8 h—48%, after 20 h—42%, and after 45–50 h—23%. When there is a high algae content, the filter holes become clogged so quickly, that there is barely enough purified water to wash them. In this case, significant pressure losses lead to reduced filter cycles and an increase in the number of filter rinsing [21, 22]. To remove algae, it is recommended to pre-treat the water with algicides, activated carbon or coagulants before water intake. To remove odours forced aeration of water and the use of special microfilters before the filtration process are recommended. Microfilters can reduce the number of filter rinsings and increase the continuity of their operation from 14 to 52 h [23].

If eutrophicated sources are used for drinking water supply it is recommended to arrange water intake in deep water layers or change their level depending on the degree of algae development and water quality. Technological schemes for the treatment of eutrophicated water can be implemented in various systems. Water containing a lot of algae should be treated in special technological systems consisting of the following stages:

- preliminary oxidation, where the most effective reagent is chlorine; if its use is not possible (due to the formation of THM), pre-ozonation of water will be a good solution;
- coagulation with an appropriately selected dose of coagulant, which allows to achieve minimal turbidity and zero zeta potential;
- sedimentation in sedimentation tanks with a suspended sludge layer or use of pressure flotation;
- filtration using sand filters or two-layer beds.

In order to achieve greater efficiency of microalgae removal, ozonation and filtering with granulated activated carbon can be additionally introduced, but their complete removal can only be achieved using membrane technologies. Technological scheme of surface water treatment with periodic phytoplankton blooms is shown in Fig. 3.10.

The efficiency of algae removal in various technological systems of water purification is presented in Table 3.5.

The efficient technology for water purification with a high content of plant biomass, *Cristal Etendu*, has proven itself well in water treatment plants in the cities of Flers, Avranches, Figny, Anders and on the island of Martinique and is presented in Fig. 3.11.

Moreover, the mass growth of algae leads to problems related to the water supply of hydropower plants for technical purposes, because clogging of filters, contamination of capacitors and pipelines reduces the reliability and efficiency of power plant operation. For example, during the period of intense blooms, when the concentration of suspensions in water for cooling purposes increases, suspensions in the amount of 3–6 Mg/day may precipitate in the pipelines of cooling circuits, which leads to the formation of deposits and reduced thermal conductivity [25].

Changes of oxidation–reduction conditions and gas regime also limits safe water use. Oxygen deficits, the development of anaerobic conditions and the associated appearance of free carbonic acid, hydrogen sulphide, ammonia and other compounds deteriorate the quality of water in the hypolimnion and consequently lead to corrosion of the pipes and hydrotechnical structures. The corrosive properties of water depending on the concentration of certain chemical compounds are presented in Table 3.6.

3.2.2 *Wastewater Treatment*

In order to prevent the intensification of eutrophication processes, the requirements regarding the content of nutrients in wastewaters discharged into water recipients are becoming more and more stringent. Modern wastewater treatment technologies make it possible to achieve very high efficiency of nitrogen and phosphorus removal in the

Fig. 3.10 Basic water purification technology for eutrophicated waters

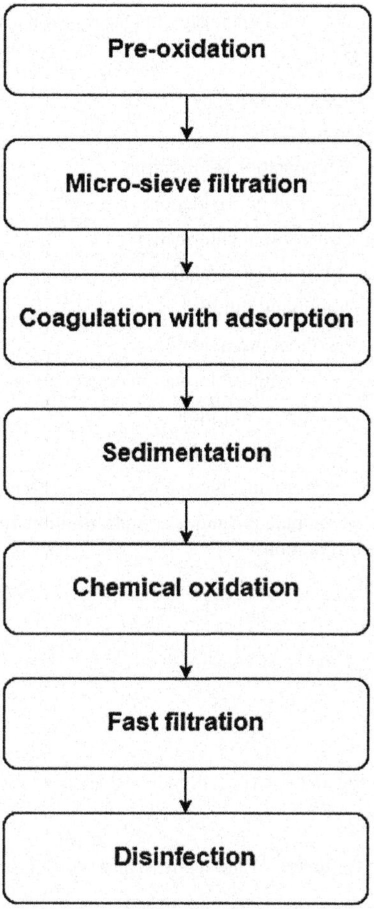

systems with enhanced biological nutrients removal (EBNR) which complies with currently applicable regulations. The efficiency of existing widely used treatment technologies is shown in Table 3.7.

In practice, the issue of using modern wastewater treatment technologies is associated with many problems, primarily of a financial nature. Conventional biological treatment plants do not provide the currently required degree of reduction of biogenic compounds. Only modern three-stage treatment plants or modernization of old ones can meet the increased quality standards of treated wastewater. The societies cannot afford almost 100% removal of pollutants from wastewater, which involves great additional costs. An illustration of dependence of treatment expenses on the treatment level was presented by Jiang et al. [28] on the example of simulation of the costs of increasing the degree of total phosphorus reduction when modernizing a biological treatment plant with simultaneous precipitation of phosphorus using $Al_2(SO_4)_3$ [28].

Table 3.5 Efficiency of algae removal (on the base of [18])

Technology	Removal efficiency %
Micro-perlocation	40–70
Direct filtration	90–98
Sedimentation + filtration	
– without pre-oxidation	95–99
– after initial chlorination	99–99.9
– after initial ozonation	97–99.8
Flotation + filtration	
– without pre-oxidation	96–99
– after initial chlorination or ozonation	99–99.9
Flotation + filtration	
– without pre-oxidation	99–99.9
– after initial chlorination or ozonation	
Membrane techniques (microfiltration, ultrafiltration after sedimentation or flotation)	>99.99

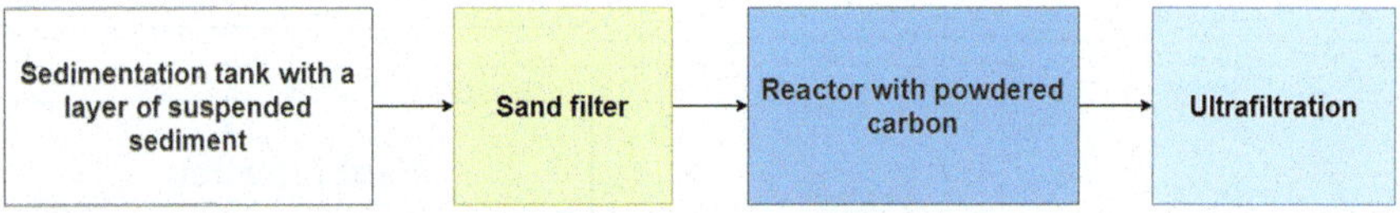

Fig. 3.11 Cristal Etendu technology of water treatment [24]

Table 3.6 Water aggressiveness according to the French standard FD P 18.011 [26]

Water aggressiveness class	A 1	A 2	A 3	A 4
Environment	Slightly aggressive	Moderately aggressive	Strongly aggressive	Extremely aggressive
Aggressiveness factors	Concentration mg/dm^3			
CO_2 aggressive	15–30	30–60	60–100	>100
SO_4^{2-}	250–600	600–1500	1500–6000	>6000
NH_4^+	15–30	30–60	60–100	>100
pH	6.5–5.5	5.5–4.5	4.5–4.0	<4.0

When calculating the unit costs of removing 1 kg of total P with various configurations of the technological schemes, expenditure on energy, labour, chemical reagents, sludge disposal, operating costs and equipment maintenance were taken into account (Table 3.8).

Table 3.7 Nutrients removal efficiency in ENBR technologies (on the base of [27])

Applied processes	Applied facilities	Efficiency (%)
High biological dephosphotation	Anaerobic–aerobic biological reactors	TN—20–40 TP—80–95 (1–5 gP/m^3)
Nitrification + denitrification	Anoxic-aerobic biological reactors for nitrogen removal	TN—70–85 TP—40–50
High biological dephosphotation + Nitrification + denitrification	Anaerobic-anoxic-aerobic biological reactors	TN—70–85 TP—80–95
Nitrification + denitrification + chemical precipitation	Anoxic–aerobic biological reactors + chemical phosphorus precipitation	TN—70–85 NP—80–95

Table 3.8 Unit costs of biological phosphorus removal depending on the efficiency of the treatment plant, degree of reduction and selected technology (based on [28])

Target TP concentration in treated wastewater		Unit costs ($/kg TP)				
		Efficiency of the treatment plant m^3/day				
		3785	*37,850*	*75,700*	*189,250*	*378,500*
2.0 mg/l	AS + Al	30.8	10.2	8.98	8.20	7.91
	A/O	61.2	27.2	23.4	20.0	18.7
	A/A/O	72.6	29.0	24.5	20.6	19.5
1.0 mg/l	AS + Al	28.4	14.9	14.1	13.5	13.3
	A/O + Al	68.0	35.4	31.5	28.3	26.9
	A/A/O + Al	87.0	37.7	32.9	29.1	27.6
0.5 mg/l	AS + Al + F	195.0	71.6	61.4	55.8	54.5
	A/O + Al + F	237.0	87.4	74.8	66.8	64.4
	A/A/O + Al + F	269.0	89.0	75.8	67.1	65.1
0.13 mg/l	AS + Al + F	250.0	138.0	129.0	125.0	123.0
	A/O + Al + F	268.0	130.0	118.0	111.0	109.0
	A/A/O + Al + F	277.0	131.0	119.0	111.0	110.0
0.05 mg/l	AS + Al + S + F + UF	387.0	200.0	183.0	167.0	162.0
	A/O + S + Al + F + UF	416.0	207.0	187.0	169.0	163.0
	A/A/O + Al + S + F + UF	423.0	206.0	186.0	168.0	162.0

AS + Al activated sludge reactor + introduction of $Al_2(SO_4)_3$; *A/O* anoxic-aerobic reactor; *A/A/O* anoxic-anaerobic–aerobic reactor; *F* filter; *S* intermediate clarifier; *UF* ultrafiltration

The simulation results are presented graphically at Fig. 3.12 from which it follows, that as the degree of pollutants reduction increases, the treatment costs increase almost exponentially.

It can be concluded that the limitation of the nutrients loads from the main source of their supply, which is wastewaters, is associated not only with technological tasks, but

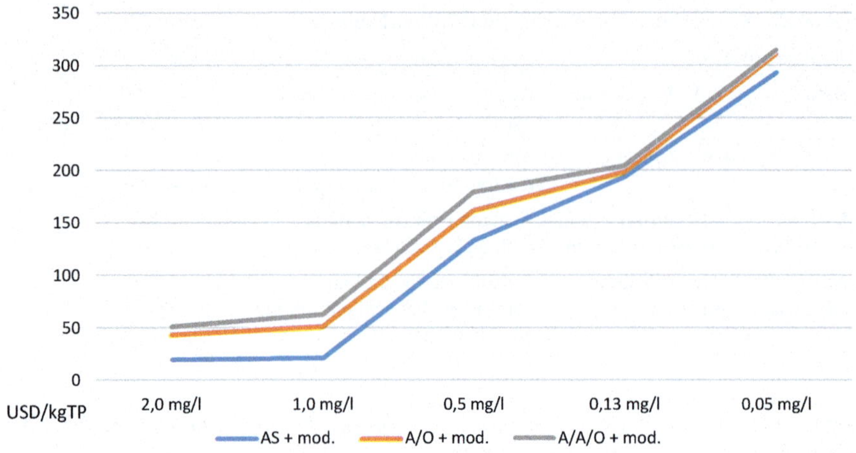

Fig. 3.12 Unit costs of removing total phosphorus in a WWTP with the flow ~17,500 m^3 per day: AS—only aerobic chamber, A/O—anoxic-aerobic chambers, AAO—anaerobic-anoxic-aerobic chambers

also results in a significant increase in investment and operational costs of wastewater treatment plants.

3.3 Economical Losses

The costs related to the development of surface water eutrophication, as well as all costs resulting from anthropogenic pressure on natural resources, may be considered in categories of environmental costs and resource costs. Environmental costs are the costs of damage caused by the water use and losses incurred by consumes of the deteriorated water environment. Resource costs are the costs of lost benefits that consumers could have achieved, if the resources had not been depleted and degraded [29]. Assessing economic damage due to eutrophication is quite difficult due to very many negative direct and indirect consequences affecting various sectors of the national economy, as well as ecological, health and social consequences. The consumers, directly or potentially impacted by negative consequences of eutrophication include: municipal and private water supply, households; recreation (swimming, boating); tourism-related industries; aquatic life, including commercial fisheries; agricultural and industrial water supply and navigation [30].

Economic losses due to eutrophication are conditioned, among others, by an increase in:

- costs of water treatment to remove phosphorus and nitrogen compounds, products of the organic substances decomposition and algae toxins,
- costs of wastewater treatment for increased removal of nutrients,

- costs of lakes, reservoirs and watercourses reclamation,
- losses of recreational and tourist value of surface waters and coastal zones,
- expenditure on water-born diseases treatment,
- losses in fishing,
- ecological damage that is difficult to quantify due to the negative impact on the aquatic biocenosis and the general disturbance of the ecological balance of water ecosystems.

Financial losses related to eutrophication are presented at Fig. 3.13.

The combined costs of annual value losses in recreational water usage, waterfront real estate, spending on recovery of threatened and endangered species and drinking water were approximately $2.2 billion annually as a result of eutrophication in U.S. freshwaters. The greatest economic losses were attributed to lakefront property values ($0.3–2.8 billion per year, although this number was poorly constrained) and recreational use ($0.37–1.16 billion per year) [30]. Figure 3.14 presents the percent increase in hypereutrophic status for lakes, reservoirs and ponds, and the combined annual fishing and boating value loss with 3 or 1 month summer closure period for each ecoregion.

For example, the Cadmus Group Inc., Texas estimated the total costs incurred from 2002 through 2012 to address poor drinking water quality due to excess nutrients for Ohio and Texas [32]. They estimated, that the City of Waco potentially lost up to

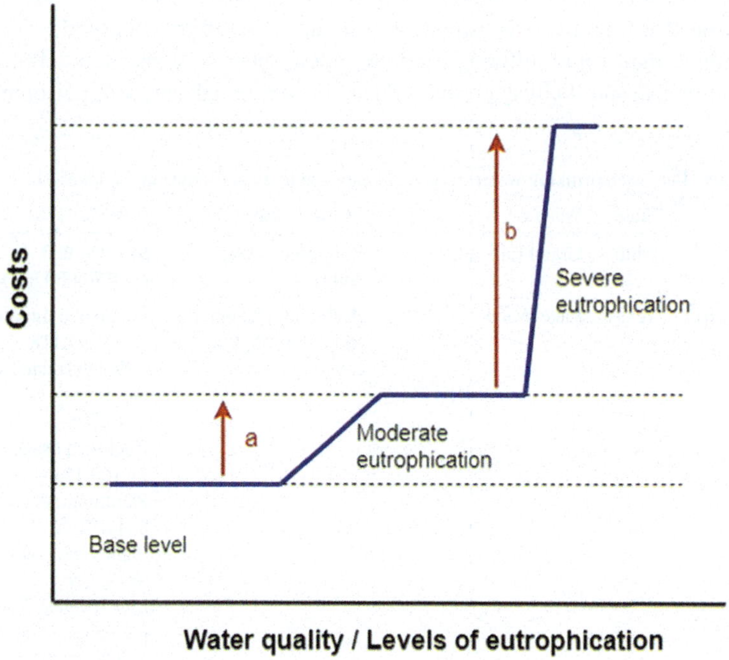

Fig. 3.13 Environmental costs and resource costs of eutrophication (after Graham et al. [31])

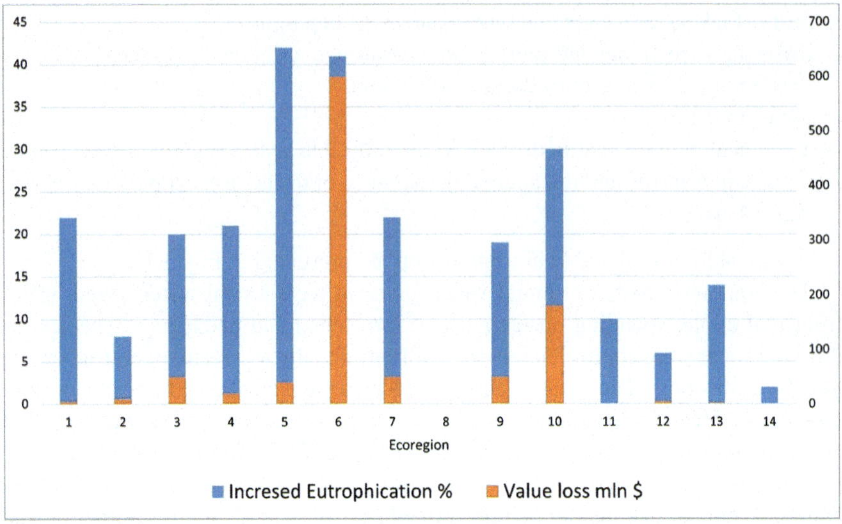

Fig. 3.14 Value loss in fishing and boating due to increased eutrophication [30]

$10.3 million in revenue due to taste and odor problems resulting in decreased water sales to neighbouring communities (Table 3.9).

The annual economic costs related to illness and death caused by consumption of poisoned fish products is estimated at around 30 million US$ worldwide [34]. The fishing industry is suffering economic losses because consumers refrain from consuming fish and shellfish products due to the perceived danger of poisoning. For

Table 3.9 Increased drinking water treatment costs due to algal blooms in the lakes [33]

Date	State	Waters	Water quality	Costs (2012$)
2010	Ohio	Grand Lake St. Marys	Blue-green algae outbreak	$13,080,000 ($3,570,000 in O&M)
2002–2012	Texas	Lake Waco	High total phosphorus and chlorophyll-a concentrations	Watershed monitoring = $2,597,118 Influent/Treated water monitoring = $740,705 Chemical usage = $1,169,151 Plant upgrades = $64,877,721 Plant energy costs = $812,755 Lost revenue from purchased water = $10,300,000

the USA, the loss related to this avoidance behavior is estimated at around 1 billion US$ per year [35].

Nutrient pollution from livestock waste impacts both fresh and marine coastal waters. The analysis reveals that every excess kilogram of phosphorus runoff from livestock waste results in total economic losses of 74.5 USD [36].

The damage costs of freshwater eutrophication in England and Wales was estimated to be \$105–160 million year^{-1}. The costs considered the reduced value of waterfront dwellings, drinking water treatment costs for nitrogen removal, reduced recreational and attractiveness value of water bodies, drinking water treatment, costs for removal of algal toxins and decomposition products, reduced value of non-polluted atmosphere, negative ecological effects on biota, and net economic losses from the tourist industry [37].

The Baltic Sea harbours one of the largest dead zones in the world. Lost revenues from ecosystem services (fishing, recreation, but also non-use values) as a result of eutrophication amount to 3–4 billion euros according to HELCOM [38]. HELCOM Executive Secretary states, that up to 4.4 billion euros are lost every year across all sectors in Baltic Sea region due to a variety of effects, that include reduced water clarity, increases in blue-green algal blooms, damage to underwater meadows, changes of the composition of fish species and oxygen deficiency on the seafloor [39]. Annual losses of benefits in the Baltic Sea region are presented at Fig. 3.15.

The economic damage costs of eutrophication estimated for the Canadian side of Lake Erie under a business as usual scenario, for example, at CAD \$270 million per year. The estimated least-cost way to reduce TP emissions by 40% in all Great Lake amounts to a total annual cost of 3 billion Canadian dollars or 0.15% of Canada's GDP [41].

The Gulf of Gabes (Lesser Syrtis Gulf) in the Mediterranean Sea, just off the south-east coast of Tunisia, used to be the location of the most extensive seagrass bed in the Mediterranean basin, now about 90% of the seagrass ecosystem surface

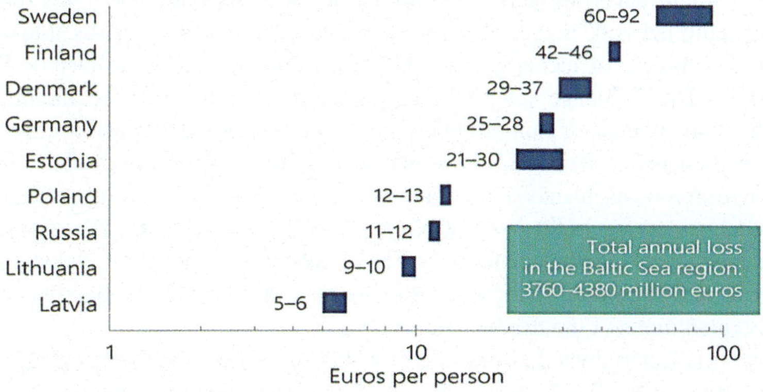

Fig. 3.15 Annual loss of benefits in the countries of Baltic Sea catchment area [40]

area is lost. The analysis estimated economic losses from loss of ecosystem services at 105 million in 2014 [42].

Deterioration of water quality as a result of eutrophication leads to a loss of recreational values and, consequently, a decrease in the cost of real estate near eutrophicated water reservoirs. Therefore, improving water quality, expressed by improving only one of the eutrophication indicators, namely increasing its transparency, may contribute to real estate costs growth and reducing losses in this economic sector [43]. For example, near the lakes in the US state of Main, an improvement in water transparency by 1 m increased the value of 1 m^2 of property located on the shores of the lakes went up by US$34–620, depending on the location in the lake catchment [44].

3.4 Eutrophication Versus Climate Change

Currently, we are observing not only the transformation of separate water bodies, but a progressive changes in the structure of the hydrosphere as a whole, possibly caused by the changes in ocean circulation and river flow volume associated with global climate change and anthropogenic activities. Oceanic biogeochemical processes are mostly determined by the structure of the phytoplankton community. Almost half of the fixed carbon on the planet comes from phytoplankton, with its biomass accounting for about 0.2% of the total biomass of all plants on the planet [45]. Phytoplankton plays a significant role in the functioning of the planet's ecosystem by regulating climate. In turn, climate affects the structure and functioning of phytoplankton [46].

Therefore, an increase in the temperature of the aquatic environment is in many cases accompanied by a noticeable increase in algae biomass. An increase in temperature can also stimulate the development of higher aquatic vegetation (macrophytes) [47]. The so-called "thermal eutrophication" which consists in the entry of additional amounts of nutrients (nitrogen, phosphorus, etc.) into the water due to the intensive hydrographic network is one of the main climate-forming factors. Its transformation in the process of technogenesis can affect climatic and hydrometeorological conditions. This influence is both direct and indirect. In the first case, the impact is carried out in the form of changes in humidity conditions, transfer and accumulation of thermal energy contained in the water mass and intensification of the biological cycle in reservoirs at elevated temperatures. As a result, heating of water often causes a noticeable increase in the level of bioproduction processes [48]. Negative consequences are caused by the occurrence of an imbalance of production-decomposition processes, which consists in the underutilization of the part of created organic matter in subsequent links of the trophic chains.

River flow determines the structure of water masses and the hydrological regime of large areas of the World Ocean, which in turn determines climatic and hydrometeorological conditions on an even larger scale. Historical changes in the circulation system of the World Ocean, provoked by the development of the greenhouse effect, have already caused an increase in the frequency and scale of hydrometeorological fluctuations. Almost each of these phenomena has significant environmental consequences on a regional and interregional scale. An even greater danger is posed by sustainable climate changes caused by technogenic transformation of the hydrological structure of the World Ocean. They will inevitably be accompanied by a radical transformation of not only aquatic, but also terrestrial ecosystems in vast areas of the planet (climatogenic successions), which will entail the need to change the nature of environmental management (ecosystem services) and a change in traditional forms of agriculture. These processes are currently observed in the form of desertification of some regions.

In this regard, some paleoecological materials are of great interest, which, for example, indicate that 7–8 thousand years ago, during the period of climate warming, natural eutrophication of lakes in Switzerland was observed [49]. At the present stage, similar phenomena may arise as a result of global warming [50]. The term "climatogenic eutrophication" can be proposed to denote them. Since this phenomenon is caused by the development of the greenhouse effect, it can be considered as a consequence of climate changes.

The water column of many bodies of water—from relatively small continental ones, such as lakes and reservoirs, to the oceans—is permanently or temporarily stratified. At the same time, in most stratified reservoirs, due to various reasons (for example, as a result of seasonal temperature changes), mixing of waters of different layers (mixia) periodically occurs. Taken together, the nature of the stratification of water masses and a certain periodicity of mictic processes constitute the stratification regime of a particular reservoirs. At the same time, the nature of the stratification of water masses composing specific reservoirs is very conservative, since it is mainly determined by such weakly changing factors as climate and geomorphological features of these reservoirs. In this regard, any violations of the stratification regime have a noticeable impact on the ecological state of the reservoir.

For example, the enrichment of surface waters with compounds of nitrogen, phosphorus and other nutrients (i.e., destratification eutrophication) that occurs as a result of disruption of stratification, can cause intense "blooming" of phytoplankton [51, 52]. In marine waters, as a result of such phenomena, the occurrence of "red tides" is possible. These phenomena are often accompanied not only by mass deaths of aquatic organisms poisoned by algae secretions, but also pose a serious danger to human health when swimming [53]. Enrichment of the surface layer with nutrients from the deeper horizons of the reservoir can also cause a significant increase in the growth of some macrophytes [54]. However, destratification of the aquatic environment, like any other significant change in the living conditions for organisms, leads to the destruction of the structural and functional organization of the pelagic ecosystem. In particular, it disrupts the balance of production and destruction processes in technogenesis of water bodies.

It should be noted that when reaching a sufficiently large scale, destratification eutrophication and the resulting imbalance of production and destruction processes can affect the course of the biogeochemical carbon cycle and thereby cause climate change [54]. A number of paleontological and geological data indicate, that it was large-scale disturbances in the stratification of sea basins, that arose as a result of various cataclysms that caused changes in the global climate in the geological past [55].

The process of eutrophication of water bodies has acquired such a global scale in recent decades, that today we are talking about lakes as "hot spots" of the carbon budget of the biosphere [56]. In connection with modern climate processes, attention is drawn to the problem of carbon dioxide emission from water bodies. Many research conducted in recent years has focused on freshwaters as a source of greenhouse gas emissions into the atmosphere. The scientists from University of Minnesota state that "eutrophication of the world's lakes will increase the emission of methane into the atmosphere by 30–90% during the next 100 years" [57]. A new study on GHG from Africa allowed concluding that lakes were significant methane sources [58]. Scientists from Finland and the USA found, that boreal lakes in Finland are responsible for approximately 30% of biogenic methane emissions increased under changing climate conditions and will increase by 26–59% by the end of this century [59]. The authors from University of Wisconsin-Madison assume that "freshwater ecosystems account for half of global emissions of methane" [60]. Figure 3.16 illustrates the main pathways of methane emissions from a ponds: diffusion, ebullition, and plant-mediate transportation [61].

Research on greenhouse gas emissions from dam reservoirs deserves particular attention. Over the past few decades, this issue has been a hot topic. Particular attention to dam reservoirs as sources of greenhouse gases is due to the fact, that they retain organic carbon carried out by rivers. In reservoirs, organic matter decomposes within a short time, and greenhouse gases are released, due to which the cycle of greenhouse gases sharply accelerates (Table 3.10).

Other researches show the correlation between water productivity and methane emissions from the reservoir surface. Chlorophyll-a and air temperature were significant predictors of CH_4 emissions from water reservoir surfaces [63].

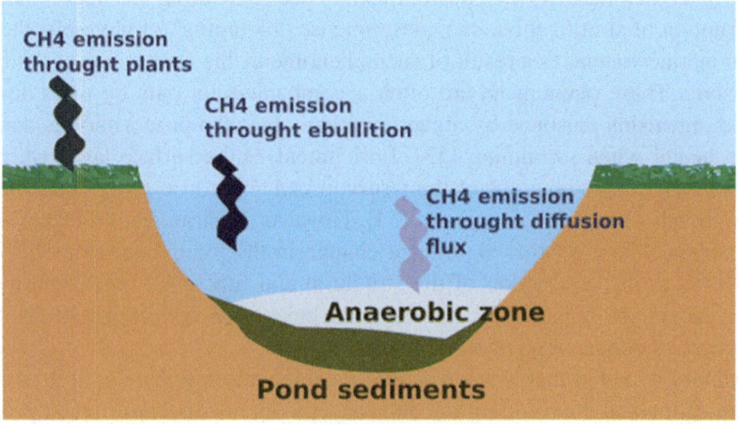

Fig. 3.16 Mechanisms of methane emissions from the pond [61]

Table 3.10 Greenhouse gas emissions from reservoirs of different climatic zones [62]

Zone	Mg CO_2 m^{-2} day^{-1}			Number of reservoirs	mg CH_4 m^{-2} day^{-1}			Number of reservoirs
	Average	Min.	Max.	n	Average	Min.	Max.	n
Moderately cold	1870	85	5750	50	17	−5	113	39
Moderate	550	−1190	4980	43	9	3	21	10
Subtropical	780	−1180	4790	36	7.7	4.2	10	5
Tropical	4000	−860	10,400	20	137	−137	1140	22

Figure 3.17 presents the simplified illustration of complicated interplay between eutrophication and temperature increase.

Marine phytoplankton fixes half of the carbon bound during photosynthesis from atmospheric carbon dioxide (about 30 and 50 billion metric tons of carbon per year) and releases half of the total volume of oxygen returned to the atmosphere despite the fact, that its total biomass is only 1% of the total plant biomass [65]. In this regard, phytoplankton quickly responds to global climate change.

Increased solar radiation and temperature, desalination of the upper layers of water enhances stratification and reduces the flow of nutrients from deep layers to the surface, which brings to the productivity decreasing. On the contrary, an increase in carbon dioxide content can increase phytoplankton productivity provided under availability of nutrients [66, 67].

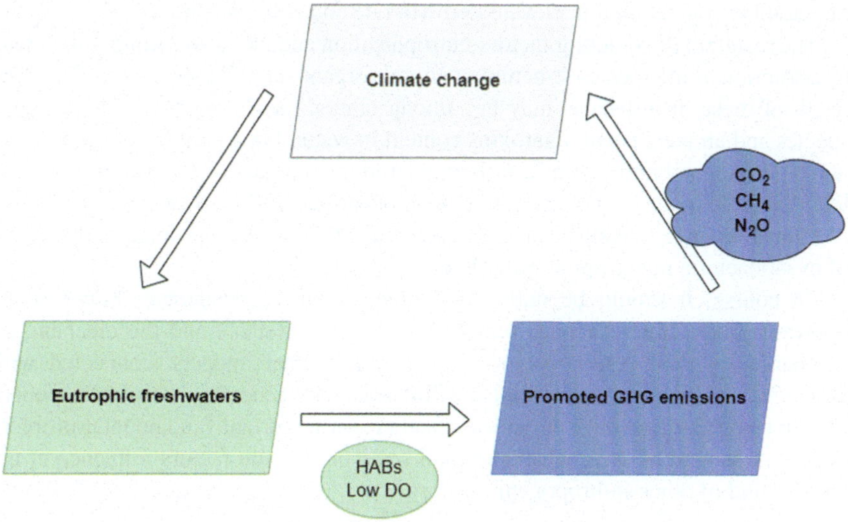

Fig. 3.17 Mutual inverse responses between eutrophication and climatic conditions [64]

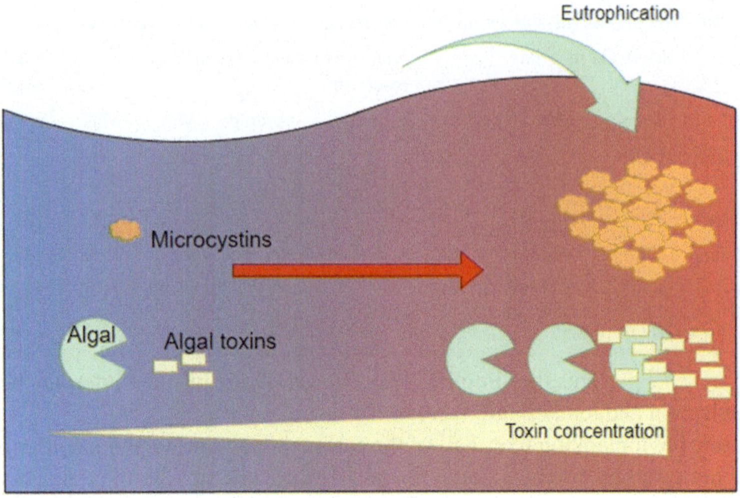

Fig. 3.18 The resultant of co-acting factors [69]

A number of studies have shown the effect of increased carbon dioxide content on certain types of phytoplankton, which can disrupt ecosystem processes and selectively stimulates toxic microscopic algae *Vicicitus globosus,* which disrupts the distribution of biomass between trophic levels of water ecosystem [68]. Marine cyanobacteria *Trichodesmium* responds to long-term (4.5 years) stay in conditions of increased carbon dioxide, which causes irreversible genetic changes in them manifesting in increased nitrogen fixation, elevated growth rate and cell size [68].

The resultant of co-acting factors-eutrophication and global warming—may lead to disruption of the species structure of aquatic organisms and their functioning. The result of these disturbances may be, among others, the dominance of toxic algae species and an increase of algatoxins content in water. For example, eutrophication and higher temperatures significantly increased the biomass of *Microcystis*, leading to an increase in the cyanobacterial toxin concentrations (Fig. 3.18).

Climate change is likely to increase the frequency, intensity and duration of blooms of cyanobacteria in eutrophic water bodies [70, 71].

Of course, it should be understood, that the interdependencies between the processes controlling the functioning of aquatic ecosystems and the mechanisms in charge for their responses to external and internal impacts, connected with climate destabilization, are not linear. Climate change generates complex responses in water bodies depending on various factors. An important fundamental problem is the assessment of the contribution and scale of climatic factors influence at the background of other anthropogenic factors.

References

1. Burns NM, Rutherford JC, Clayton JS (1999) A monitoring and classification system for New Zealand lakes and reservoirs. Lake Reservoir Manag 15(4):255–271. https://doi.org/10.1080/07438149909354122
2. Caille C (1948) Die künstliche Erzeugung der Wasserzirkulation in Seen. SIL Proc 1922–2010 10(1):104–105. https://doi.org/10.1080/03680770.1948.11895124
3. Carlson RE (1977) A trophic state index for lakes. Limnol Oceanogr 22(2):361–369. https://doi.org/10.4319/lo.1977.22.2.0361
4. Neverova-Dziopak E (2021) Surface water eutrophication in Poland: assessment and prevention. In: Quality of water resources in Poland. Springer, pp 321–346. https://doi.org/10.1007/978-3-030-64892-3_14
5. UN (2010) Resolution 64/292. The human right to water and sanitation. In: Treaty series, vol I, issue 9464
6. Sirenko L, Gawrilenko V (1978) Cwietienie wody i ewtrofirowanie. Naukowa Dumka
7. Zhang W, Liu J, Xiao Y, Zhang Y, Yu Y, Zheng Z, Liu Y, Li Q (2022) The impact of cyanobacteria blooms on the aquatic environment and human health. Toxins 14(10):658. https://doi.org/10.3390/toxins14100658
8. Kubickova B, Babica P, Hilscherová K, Šindlerová L (2019) Effects of cyanobacterial toxins on the human gastrointestinal tract and the mucosal innate immune system. Environ Sci Eur 31(1):31. https://doi.org/10.1186/s12302-019-0212-2
9. Paul V, Shamah S, Garankina O, Wolf L, Shia Lin Y (2014) Rhabdomyolysis after fish consumption: Haff's disease. QJM 107(1):67–68. https://doi.org/10.1093/qjmed/hct172
10. Siwkow GS, Razmaszkin GA, Listoszenko AL, Kapustna JA (2003) SibRybNII Projekt. Materiały WNII Weterynarnoj Entomologii
11. Backer L, Manassaram-Baptiste D, LePrell R, Bolton B (2015) Cyanobacteria and algae blooms: review of health and environmental data from the harmful algal bloom-related illness surveillance system (HABISS) 2007–2011. Toxins 7(4):1048–1064. https://doi.org/10.3390/toxins7041048
12. EC (2002) Eutrophication and health
13. Johnson PTJ, Chase JM, Dosch KL, Hartson RB, Gross JA, Larson DJ, Sutherland DR, Carpenter SR (2007) Aquatic eutrophication promotes pathogenic infection in amphibians. Proc Natl Acad Sci 104(40):15781–15786. https://doi.org/10.1073/pnas.0707763104
14. CDC (2019) Summary report—One Health Harmful Algal Bloom System (OHHABS)
15. Lee S, Choi B, Kim SJ, Kim J, Kang D, Lee J (2022) Relationship between freshwater harmful algal blooms and neurodegenerative disease incidence rates in South Korea. Environ Health 21(1):116. https://doi.org/10.1186/s12940-022-00935-y
16. Kirpenko J, Sirenko L, Orlowski W, Bokow A (1975) Sodierżanie toksyczeskich metabolitow siniezielonych wodoroslej w wodzie wo wremia „cwietienia" i ich wlijanie na organizm teplokrownych ziwotnych. Samooczyszczenie, biopruduktiwnost i ochrana wodojemow i wodotokow Ukrainy
17. Schwimmer D, Schwimmer M (1964) Alga and medicine. Alga and man. Jackson D.T.N.W.
18. Chaussade J-L (2005) Memento technique de l'eau. Tome 1, 10th edn. Degrémont SA, Rueil-Malmaison
19. Lund JWG (1966) Limnology and its application to potable water supplies. J Brit Waterworks Assoc 49
20. Choma J, Jaroniec M, Burakiewicz-Mortka W, Gwizdalski M (1995) Adsorpcja związków organicznych z roztworów wodnych na węglach aktywnych. Ochrona Środowiska 1(56)
21. Liu W, Qiu R (2007) Water eutrophication in China and the combating strategies. J Chem Technol Biotechnol 82(9):781–786. https://doi.org/10.1002/jctb.1755
22. Tsvetkova L, Neverova-Dziopak E, Usanov B (2000) K woprosu ob kologiczeskom monitoringie wodnych korridorow. Puti Rieszenia Ekologiczeskich Problem Transportnych Korridorow

23. Tsvetkova L, Neverova-Dziopak E, Aleksejew M, Basowa S, Tronin A, Rumin G, Silin B (2006) Issledowanie wlijania stocznych wod Sankt Petersburga na ewtrofirowanie i wtoricznoje zagriaznienie newsko guby i wostocznoj czasti Finskogo Zaliwa. SPbGASU
24. Degremont (2005) Memento technique de l'eau, 2nd edn. Rueil-Malmaison
25. Kanygina A, Wanin W (1974) Rekomendacji po eksploatacji wodojemow-ochladitielejb sistem oborotnogo wodosnabżenija i predotwraszenia ich ewtrofikacji
26. AFNOR (1992) Bétons. Classification des environnements agressifs NF P 18.011
27. Dymaczewski Z (2006) Wodociągi i Kanalizacja w Polsce – tradycja i współczesność. Polska Fundacja Ochrony Zasobów Wodnych
28. De Novo C, Jiang F, Rowles KL, Russell D (2004) Estimation of costs of phosphorus removal in wastewater treatment facilities. https://api.semanticscholar.org/CorpusID:31581908
29. EU (2002) Common implementation strategy for the water framework directive (2000/60/EC)
30. Dodds WK, Bouska WW, Eitzmann JL, Pilger TJ, Pitts KL, Riley AJ, Schloesser JT, Thornbrugh DJ (2009) Eutrophication of U.S. freshwaters: analysis of potential economic damages. Environ Sci Technol 43(1):12–19. https://doi.org/10.1021/es801217q
31. Graham M, Blignaut J, Villiers L, Mostert D, Sibande X, Gebremedhin S, Harding B, Rossouw N, Freese N, Ferrer S, Browne M (2012) Development of a generic model to assess the costs associated with eutrophication
32. Davenport T, Drake W (2011) Ohio—the case for source water protection. In: EPA nutrients and algae blooms
33. U.S EPA (2009) A compilation of cost data associated with the impacts and control of nutrient pollution
34. Kouakou CRC, Poder TG (2019) Economic impact of harmful algal blooms on human health: a systematic review. J Water Health 17(4):499–516. https://doi.org/10.2166/wh.2019.064
35. Parsons GR, Morgan A, Whitehead JC, Haab TC (2006) The welfare effects of Pfiesteria-related fish kills: a contingent behavior analysis of seafood consumers. Agric Resour Econ Rev 35(2):348–356. https://doi.org/10.1017/S106828050000678X
36. Sampat AM, Hicks A, Ruiz-Mercado GJ, Zavala VM (2021) Valuing economic impact reductions of nutrient pollution from livestock waste. Resour Conserv Recycl 164:105199. https://doi.org/10.1016/j.resconrec.2020.105199
37. Pretty JN, Mason CF, Nedwell DB, Hine RE, Leaf S, Dils R (2003) Environmental costs of freshwater eutrophication in England and Wales. Environ Sci Technol 37(2):201–208. https://doi.org/10.1021/es020793k
38. Hyytiäinen K, Blyh K, Hasler B, Ahlvik L, Ahtiainen H, Artell J, Ericsdotter S (2014) Environmental economic research as a tool in the protection of the Baltic Sea—costs and benefits of reducing eutrophication. https://doi.org/10.6027/TN2014-504
39. Strempel R (2022) The cost of degradation: a price tag on the Baltic Sea. Baltic Rim Econ 1
40. HELCOM (2023) State of the Baltic Sea 2023. HOLAS 3
41. Garcia-Hernandez JA, Brouwer R, Pinto R (2022) Estimating the total economic costs of nutrient emission reduction policies to halt eutrophication in the Great Lakes. Water Resour Res 58(4). https://doi.org/10.1029/2021WR030772
42. El Zrelli R, Hcine A, Yacoubi L, Roa-Ureta RH, Gallai N, Castet S, Grégoire M, Courjault-Radé P, Rabaoui LJ (2023) Economic losses related to the reduction of Posidonia ecosystem services in the Gulf of Gabes (Southern Mediterranean Sea). Mar Pollut Bull 186:114418. https://doi.org/10.1016/j.marpolbul.2022.114418
43. Middlebrooks EJ, Pearson EA, Tunzi M, Adinarayana A, McGauhey PH, Rohlich GA (1971) Eutrophication of surface water: Lake Tahoe. J Water Pollut Control Fed 43(2):242–251. http://www.jstor.org/stable/25036890
44. Michael H, Boyle K, Bouchard R (1998) Water quality affects property prices: a case study of selected Maine lakes. Maine Agr. Forest Exp. Station misc report, vol 398
45. Field CB, Behrenfeld MJ, Randerson JT, Falkowski P (1998) Primary production of the biosphere: integrating terrestrial and oceanic components. Science 281(5374):237–240. https://doi.org/10.1126/science.281.5374.237

46. Falkowski PG, Oliver MJ (2007) Mix and match: how climate selects phytoplankton. Nat Rev Microbiol 5(10):813–819. https://doi.org/10.1038/nrmicro1751
47. Beznosov V, Kuchkina M, Suzdaleva A (2002) Study of the process of thermal eutrophication in cooling ponds of nuclear power plants. Water Resour 29(5)
48. Suzdaleva A, Goryunova S (2014) Technogenesis and degradation of surface water objects. OOO ID ENERGIYA
49. Haas M, Baumann F, Castella D, Haghipour N, Reusch A, Strasser M, Eglinton TI, Dubois N (2019) Roman-driven cultural eutrophication of Lake Murten, Switzerland. Earth Planet Sci Lett 505:110–117. https://doi.org/10.1016/j.epsl.2018.10.027
50. Beznosov V, Suzdaleva A (2004) Potential changes in aquatic biota in the period of global climate warming. Water Resour 31(4)
51. Suzdaleva A (2001) The influence of circulating water masses of nuclear power plants on the distribution of bacterioplankton in cooling ponds. Water Resour 28(3)
52. Beznosov V, Suzdaleva A (2004) Possible changes in aquatic biota during global warming. Water Resour 31(4)
53. John Anderson N, Jeppesen E, Sondergaard M (2005) Ecological effects of reduced nutrient loading (oligotrophication) on lakes: an introduction. Freshwater Biology, 50:1589–1593. https://doi.org/10.1111/j.1365-2427.2005.01433.x
54. Beznosov V, Zhelezny B (2000) A critical volume of disruption of ocean stratification capable of causing a large-scale change in the balance of production-destruction processes and the biogeochemical carbon cycle. In: Ecosystem restructuring and evolution of the biosphere, vol 4
55. Suzdaleva A, Beznosov V (2020) Modern technogenesis of the world ocean: nature of processes and ecological problems, vol 2
56. Holgersson M, Baldwin CY, Chesbrough H, Bogers MLAM (2022) The forces of ecosystem evolution. Calif Manag Rev 64(3):5–23. https://doi.org/10.1177/00081256221086038
57. Beaulieu JJ, DelSontro T, Downing JA (2019) Eutrophication will increase methane emissions from lakes and impoundments during the 21st century. Nat Commun 10(1):1375. https://doi.org/10.1038/s41467-019-09100-5
58. Borges AV, Deirmendjian L, Bouillon S, Okello W, Lambert T, Roland FAE, Razanamahandry VF, Voarintsoa NRG, Darchambeau F, Kimirei IA, Descy J-P, Allen GH, Morana C (2022) Greenhouse gas emissions from African lakes are no longer a blind spot. Sci Adv 8(25). https://doi.org/10.1126/sciadv.abi8716
59. Guo M, Zhuang Q, Tan Z, Shurpali N, Juutinen S, Kortelainen P, Martikainen PJ (2020) Rising methane emissions from boreal lakes due to increasing ice-free days. Environ Res Lett 15(6):064008. https://doi.org/10.1088/1748-9326/ab8254
60. Rocher-Ros G, Stanley EH, Loken LC, Casson NJ, Raymond PA, Liu S, Amatulli G, Sponseller RA (2023) Global methane emissions from rivers and streams. Nature 621(7979):530–535. https://doi.org/10.1038/s41586-023-06344-6
61. Malyan SK, Singh O, Kumar A, Anand G, Singh R, Singh S, Yu Z, Kumar J, Fagodiya RK, Kumar A (2022) Greenhouse gases trade-off from ponds: an overview of emission process and their driving factors. Water 14(6):970. https://doi.org/10.3390/w14060970
62. Fedorov MP, Elistratov VV, Maslikov VI, Sidorenko GI, Chusov AN, Atrashenok VP, Molodtsov DV, Savvichev AS, Zinchenko AV (2015) Reservoir greenhouse gas emissions at Russian HPP. Power Technol Eng 49(1):33–39. https://doi.org/10.1007/s10749-015-0569-3
63. Deemer BR, Harrison JA, Li S, Beaulieu JJ, DelSontro T, Barros N, Bezerra-Neto JF, Powers SM, dos Santos MA, Vonk JA (2016) Greenhouse gas emissions from reservoir water surfaces: a new global synthesis. Bioscience 66(11):949–964. https://doi.org/10.1093/biosci/biw117
64. Li Y, Shang J, Zhang C, Zhang W, Niu L, Wang L, Zhang H (2021) The role of freshwater eutrophication in greenhouse gas emissions: a review. Sci Total Environ 768:144582. https://doi.org/10.1016/j.scitotenv.2020.144582
65. Behrenfeld MJ (2014) Climate-mediated dance of the plankton. Nat Clim Change 4(10):880–887. https://doi.org/10.1038/nclimate2349

66. Behrenfeld MJ, Hu Y, O'Malley RT, Boss ES, Hostetler CA, Siegel DA, Sarmiento JL, Schulien J, Hair JW, Lu X, Rodier S, Scarino AJ (2017) Annual boom–bust cycles of polar phytoplankton biomass revealed by space-based LiDAR. Nat Geosci 10(2):118–122. https://doi.org/10.1038/ngeo2861
67. Behrenfeld MJ, O'Malley RT, Siegel DA, McClain CR, Sarmiento JL, Feldman GC, Milligan AJ, Falkowski PG, Letelier RM, Boss ES (2006) Climate-driven trends in contemporary ocean productivity. Nature 444(7120):752–755. https://doi.org/10.1038/nature05317
68. Cavicchioli R, Ripple WJ, Timmis KN, Azam F, Bakken LR, Baylis M, Behrenfeld MJ, Boetius A, Boyd PW, Classen AT, Crowther TW, Danovaro R, Foreman CM, Huisman J, Hutchins DA, Jansson JK, Karl DM, Koskella B, Mark Welch DB, Webster NS et al (2019) Scientists' warning to humanity: microorganisms and climate change. Nat Rev Microbiol 17(9):569–586. https://doi.org/10.1038/s41579-019-0222-5
69. Yang Y, Wang H, Yan S, Wang T, Zhang P, Zhang H, Wang H, Hansson L-A, Xu J (2023) Chemodiversity of cyanobacterial toxins driven by future scenarios of climate warming and eutrophication. Environ Sci Technol 57(32):11767–11778. https://doi.org/10.1021/acs.est.3c02257
70. Huisman J, Codd GA, Paerl HW, Ibelings BW, Verspagen JMH, Visser PM (2018) Cyanobacterial blooms. Nat Rev Microbiol 16(8):471–483. https://doi.org/10.1038/s41579-018-0040-1
71. Paerl HW, Huisman J (2008) Blooms like it hot. Science 320(5872):57–58. https://doi.org/10.1126/science.1155398

Chapter 4
The Complexity of the Problem

Abstract This chapter presents problematic aspects of managing the eutrophication process, starting from the lack of a uniform definition of this process and discrepancies in the understanding of its essence, to various approaches to its research, mitigation and managing. Variety of widely adopted definitions of the eutrophication phenomenon were analysed and presented in the chapter. Different methodological approaches to understanding the eutrophication process were also presented: the metabolic approach and the balance approach. In this chapter the authors formulated their holistic approach and their own definition of eutrophication.

Keywords Definition of eutrophication · Metabolic approach · Holistic approach · Balance approach

4.1 Difficulties and Controversies in Defining Eutrophication

One of the major water related problem of global scale and a priority issue for water management is surface water eutrophication. At the XXII session of UNEP in 1984, the process of eutrophication of water bodies was placed in first place in terms of the degree of danger of global anthropogenic impact on the environment. Many years have passed since that time, but this problem is still far from a final solution [1].

Anthropogenic eutrophication belongs to the processes which intensity has increased significantly in recent decades and presents the most spectacular manifestation of deterioration of ecological status and functional properties of surface waters [2].

Anthropogenic eutrophication now has become the research focus not only for limnologists and hydrobiologists, but also for the specialists in the field of ecological engineering and water resource management. Many EU directives require monitoring and/or assessing eutrophication: Urban Wastewater Treatment Directive, Nitrates Directive, Water Framework Directive, Helsinki Convention and other European Directives [3].

E. Neverova-Dziopak and Z. Kowalewski, *Eutrophication: A Global Environmental Problem*, Springer Water, https://doi.org/10.1007/978-3-031-83926-9_4

The concept of eutrophication was introduced in 1907 by C. A. Weber for determining the nutrient conditions for peat bogs flora in northern Germany and considered the extinct lakes [4]. The "water trophy" concept was first introduced by German hydrobiologists A. F. Thiemann and Swedish limnologists E. C. L. Naumann. In 1921 both Thienemann and Naumann published their independent research on the biological types of lakes to describe the ability of aquatic ecosystems to produce organic substances in the process of photosynthesis. "Water trophy" was a synonym for primary production and relating physical–chemical characteristics of lakes to their biological productivity [5].

Later, this term was used to describe the general condition of aquatic ecosystems. Regardless of the fact that the term "trophy" has been widely spread, its essence and definition remain extremely undefined.

So far, there is not even a universal definition and the term eutrophication is not explicitly defined. Most of contemporary literature sources define the process of eutrophication as an enrichment of water with nutrients (biogenic substances) leading to the intensive development of aquatic vegetation and related negative consequence. According to Directive 91/271/EEC concerning urban waste water treatment (the UWWT Directive), eutrophication means "the enrichment of water by nutrients, especially of nitrogen or phosphorus compounds, causing an accelerated growth of algae and higher forms of plant life to produce an undesirable disturbance" [6]. Directive 91/676/EEC (Nitrate Directive) gives almost the identical definition of this process [7]. Water Framework Directive considers eutrophication as a process, where "anthropogenic nutrient enrichment causes adverse changes in the aquatic environment". It focuses more on changes in the aquatic environment rather than a particular state of primary production [3].

The authors of publication entitled "A new wine in an old bottle?" wrote: "The term eutrophication is used by both the scientific community and public policy-makers, and therefore has a myriad of definitions" [8].

So far, different authors have put different meanings into the term of "eutrophication" and "water trophy" (Table 4.1).

In European water management practice it was chosen to define the eutrophication in terms of nutrient pollution and "undesirable" changes in the ecological status of water bodies [15]. Nixon considers it to be an anthropocentric view of what is "desirable" in nature and the definition to be too limited, that makes eutrophication a pollution problem arising from one type of pollutant and not a complicated ecological problem [2].

As follows from the analysis of various definitions existing in literature the eutrophication is most often understood as an excessive content of nutrients in water, an increase in water productivity and its visual signs. As S. W. Nixon points out "defining eutrophication in terms of changing nutrient concentrations or chlorophyll levels or species composition confuses symptoms with the underlying phenomenon" [2].

However, no one definition listed in the Table 4.1 do not mention the disturbance of the biotic balance that occurs during this process. In these definitions, primary

Table 4.1 Variety of widely adopted definitions of the eutrophication phenomenon

Definition of eutrophication	Source
The gradual increase in the concentration of phosphorus, nitrogen, and other plant nutrients in an aging aquatic ecosystem such as a lake	Encyclopaedia Britannica [9]
The increased primary productivity (trophy) in a water body due to enhanced availability or usage of nutrients	Encyclopaedia of Ecology 2008 [10]
The process of enrichment of waters with plant nutrients, that stimulates aquatic primary production	UNEP/UNESCO/FAO [11]
An increase in the rate of supply of organic matter to an ecosystem	Eutrophication and the macroscope [2]
The natural ageing process of lakes, greatly accelerated by human intervention in the natural biogeochemical cycling of nutrients within a watershed	Trends in eutrophication research and control [12]
The excessive plant and algal growth due to the increased availability of one or more limiting growth factors needed for photosynthesis	Recent advances in the understanding and management of eutrophication [13]
The fundamental change in the energetic base that may propagate through the system in various ways and produce a variety of changes	Eutrophication and the macroscope [14]
The enrichment of water by nutrients, especially nitrogen and/or phosphorus and organic matter, causing an increased growth of algae and higher forms of plant life to produce an unacceptable deviation in structure, function and stability of organisms present in the water and to the quality of water concerned, compared to reference conditions	Coastal eutrophication: recent developments in definitions and implications for monitoring strategies [15]
The result of excessive enrichment of water with nutrients, which may accelerate the growth of algae (phytoplankton) in the water column that may result in a range of undesirable disturbances in the aquatic ecosystem	Convention for the protection of the marine environment of the north-east Atlantic [16]
The accelerated production of organic matter, particularly algae and higher forms of plant life, in a water body usually caused by an increase in the amount of nutrients being discharged to the water body	European Commission [3]

(continued)

Table 4.1 (continued)

Definition of eutrophication	Source
The process driven by enrichment of water by nutrients, especially compounds of nitrogen and/or phosphorus, leading to: increased growth, primary production and biomass of algae; changes in the balance of organisms and water quality degradation	Marine Strategy Framework Directive (2008/56/EC) (MSFD) [17]
The excessive amounts of nutrients, nitrogen (N), phosphorus (P), and sometimes organic matter (represented by carbon, C), that brings to changes in the structure and functioning of the entire marine ecosystem and a reduction in ecosystem stability	Eutrophication in the Baltic Sea [18]

production is given so much importance, that the term "trophy" becomes unnecessary because it is replaced by the primary production rate [19]. Andersen also argue strongly for including the measurement of primary production in European eutrophication monitoring programs [20].

Nevertheless, surface water ecosystem functioning is characterized not only by the rate of primary production, but also by the rate of its decomposition. The accumulation of organic substances in water depends on the ratio of these fundamental processes. So, certain scientists proposed to determine the trophic type of waters according to the state of organic matter production-decomposition balance [21]. The similar view was also expressed by Ohle, who proposed classifying lakes according to the amount of production and decomposition of phytoplankton biomass [22]. In publications of Alimov, A. F. the author also believes that the basic characteristic of aquatic ecosystems is the balance of production and decomposition processes of organic substances [23].

4.2 Balance Approach to Eutrophication

Modern aquatic ecology is mainly based on so-called metabolic approach to aquatic ecosystems, formed in the middle of the twentieth century treating water body as an integrated entirety [24]. Such approach assumes the study of ecosystem metabolism considering its functioning within the frame of two mutually opposite processes—the production of organic substances during photosynthesis and their destruction.

Already in 1887 the American zoologist Stephen A. Forbes wrote an influential essay concerning the dynamic equilibrium of water ecosystem entitled "The Lake as a Microcosm". Thienemann in his publications also used the same word, "microcosm", to describe the lake ecosystem functioning [5].

At the beginning of the twentieth century, the most outstanding scientists in the field of hydrology and limnology introduced the balance approach for understanding the ecosystems functioning. The emergence of the trophodynamic concept of ecosystem, formulated by R. Lindeman in 1942 quickly gaining strong popularity, especially in the Anglo-American scientific community [25]. Lindemann's trophodynamics is associated with taking into account the ratio of organic matter production and decomposition in water ecosystem.

The concept of water trophy assumes biological differences in water ecosystems, but so far there is no clear opinion as to which criterion may be the most adequate for the quantitative assessment of the trophic status of waters. It is very difficult to find a practically appropriate criterion, that would allow for a quantitative reflection of the complex set of conditions determining the assimilation rate of primary production, depending on abiotic and biotic factors. For this purpose, the four most common approaches were used based on: (1) biogenic substances, (2) vertical oxygen distribution during the period of temperature stratification; (3) indicator organisms (plankton, benthos); (4) primary production expressed in different units.

Rodhe W. believed that the indeterminacy of this term could be explained by the fact that the classification of water trophic state does not have a single, quantitatively measurable basis [26]. Such a basis cannot be developed as long as the trophic criteria are the conditions for primary production synthesis and the amount of this production [21].

Russian hydrobiologist L. Rossolimo considered that in the interests of solving applied tasks, the study of water ecosystems should follow the line of studying the balance of substances and energy, which is derived from the water ecological balance [27].

By 1940, G. Vinberg summarized the results of research on the quantitative patterns of metabolism of organisms and primary production of different water bodies and introduced the term "biotic balance" [28].

Surface waters are characterized not only by the rate of primary production, but also by the rate of its decomposition. Therefore, the accumulation of organic substances in water depends on the ratio of these rates. So it was proposed to determine the trophic type of water according to the type of organic matter production-decomposition balance [21].

Only in a few fundamental studies it can be found the definition, which in more substantive way reflects the essence of eutrophication process and define it as the disruption of the balance of the processes of production and decomposition of organic matter produced by algae as a result of excessive nutrient supply of surface waters, which leads to increased water productivity and its secondary pollution [29, 30].

Numerous long-term serious studies on surface water eutrophication have enabled the understanding and description of the essence of this phenomenon as the disturbance of the aquatic ecosystem functioning, caused by the violation of its biotic balance as a result of anthropogenic enrichment of waters with nutrients [30].

With the increase of the rate of water plants biomass production, the rate of its decomposition increases accordingly. The balance between the process of production and decomposition of organic matter determines the trophic status of waters [31, 32].

Table 4.2 The state of the biotic balance in waters [30]

State of biotic balance (V_p/V_d)[a]	Trophic status
Negative ($V_p/V_d < 1$)	Dystrophic/Ultraoligotrophic
Balanced ($V_p/V_d = 1$)	Oligotrophic
Positive ($V_p/V_d > 1$)	Mezotrophic/Eutrophic

[a] Ratio of of production/decomposition processes of primary organic matter

Therefore, the balance of production and destruction processes (biotic balance of autotrophic and heterotrophic processes) is one of the basic functional characteristics of any ecosystems [32]. So the trophic state and ecological state of waters in this context are synonymous.

It can be concluded, that the biotic balance seems to be the only basis on which a system of water trophic types classification can be developed. In other words, the biotic balance is a universal criterion of the trophic status of waters. Many researchers agree, that in order to quantify the biotic balance, it is necessary to establish concatenation between all processes in water bodies. But unfortunately, regardless of the justification for considering the biotic balance as one of the main trophic status criteria of aquatic ecosystems, the investigations towards its quantitative assessment did not find their development. The assessment of the trophic status is still carried out mainly on the basis of indicators, characterizing the primary production processes. Assessment of the phytoplankton photosynthesis rate and its decomposition, tested in laboratory conditions, cannot provide a reliable characterization of the entire ecosystem metabolism. The state of this balance, which is called the biotic balance, the scale of its disturbance and further consequences are determined mainly by climatic, morphometric, hydrological, hydrobiological and other characteristics of water ecosystems.

Taking into account the existing definitions of the eutrophication process, as well as the actual state of knowledge and the necessity of holistic approach to this issue, the authors propose the following definition of the term: "eutrophication is the process of gradual transition of water ecosystem from dystrophic/oligotrophic state to eutrophic state, which is based on the shift of biotic balance towards primary production dominating over its decomposition". In a very general form, three types of balance can be distinguished: negative, zero and positive (Table 4.2).

It should be emphasized that due to balance approach development in hydrobiology, a paradigm shift occurred—a transition from a biocenotic way of thinking to an ecosystem one [24].

References

1. UNEP (2007) Global environment outlook GEO environment for development 4
2. Nixon SW (2009) Eutrophication and the macroscope. Hydrobiologia 629(1):5–19. https://doi.org/10.1007/s10750-009-9759-z
3. EC (2009) Guidance document on eutrophication assessment in the context of European water policies. Guidance document No 23. Publications Office. https://doi.org/doi/10.2779/49795
4. Weber CA (1907) Aufbau und vegetation der Moore Norddeutschlands
5. Egerton FN (2014) History of ecological sciences, Part 50: Formalizing limnology, 1870s to 1920s. Bull Ecol Soc Am 95(2):131–153. https://doi.org/10.1890/0012-9623-95.2.33
6. EC (1991) Council Directive 91/271/EEC of 21 May 1991 concerning urban waste-water treatment
7. EC (1991) Council Directive 91/676/EEC of 12 December 1991 concerning the protection of waters against pollution caused by nitrates from agricultural sources
8. Le Moal M, Gascuel-Odoux C, Ménesguen A, Souchon Y, Étrillard C, Levain A, Moatar F, Pannard A, Souchu P, Lefebvre A, Pinay G (2019) Eutrophication: a new wine in an old bottle? Sci Total Environ 651:1–11. https://doi.org/10.1016/j.scitotenv.2018.09.139
9. Britannica T (2024) Eutrophication. Encyclopedia Britannica. https://Www.Britannica.Com/Science/Eutrophication
10. Fath B, Jorgensen S (2008) Encyclopedia of ecology. Elsevier
11. Aubert M (1990) Mediators of microbiological origin and eutrophication phenomena. Mar Pollut Bull 21(1):24–29. https://doi.org/10.1016/0025-326X(90)90148-2
12. Rast W, Thornton JA (1996) Trends in eutrophication research and control. Hydrol Process 10(2):295–313. https://doi.org/10.1002/(SICI)1099-1085(199602)10:2%3c295::AID-HYP360%3e3.0.CO;2-F
13. Schindler DW (2006) Recent advances in the understanding and management of eutrophication. Limnol Oceanogr 51(1part2):356–363. https://doi.org/10.4319/lo.2006.51.1_part_2.0356
14. Nixon SW (2009) Eutrophication and the macroscope. In: Eutrophication in coastal ecosystems. Springer Netherlands, pp 5–19. https://doi.org/10.1007/978-90-481-3385-7_2
15. Andersen JH, Schlüter L, Ærtebjerg G (2006) Coastal eutrophication: recent developments in definitions and implications for monitoring strategies. J Plankton Res 28(7):621–628. https://doi.org/10.1093/plankt/fbl001
16. OSPAR (2003) Convention for the protection of the marine environment of the north-east Atlantic
17. EC (2010) Marine strategy framework directive 2008/56/EC
18. HELCOM (2009) Eutrophication in the Baltic Sea
19. Babanazarova OV (1997) Phytoplankton structure, composition and production in saline waterbodies in the Lower Amu Dar'ya. Int J Salt Lake Res 6(3):217–231. https://doi.org/10.1007/BF02449926
20. Anderson N, Jeppesen E, Sondgard M (2005) Ecological effects of reduced nutrient loading (oligotrophication) on lakes: an introduction. Freshw Biol 50(10):1589–1593. https://doi.org/10.1111/j.1365-2427.2005.01433.x
21. Vinberg GG (1971) Methods for the estimation of production of aquatic animals. Academic Press
22. Ohle W (1958) Diurnal production and destruction rates of phytoplankton in lakes
23. Alimov AF (2018) Long-term dynamics of biological diversity in water basins. Biol Bull Rev 8(2):67–71. https://doi.org/10.1134/S2079086418020020
24. Rizhinashvili A (2020) N. K. Koltsov's school of experimental biology and the emergence of G. G. Vinberg's concept of the biotic balance. Voprosy Istorii Estestvoznaniia i Tekhniki 41(3):458. https://doi.org/10.31857/S020596060010868-6
25. Lindeman RL (1942) The trophic-dynamic aspect of ecology. Ecology 23(4):399–417. https://doi.org/10.2307/1930126
26. Rodhe W (1969) Crystallization of eutrophication concepts in Northern Europe. Environ Sci Geogr

27. Rossolimo J (1976) Lake accumulation of organic matter and the possibilities of its typification. Nauka
28. Vinberg GG (1963) The primary production of bodies of water. Michigan University
29. Hutchinson E (1967) A treatise on limnology, introduction to lake biology and the limnoplankton. Wiley
30. Neverova-Dziopak E (2010) Basics of managing the process of anthropogenic eutrophication. AGH
31. Neverova-Dziopak E (2021) Surface water eutrophication in Poland: assessment and prevention. In: Quality of water resources in Poland. Springer, pp 321–346. https://doi.org/10.1007/978-3-030-64892-3_14
32. Odum E (1971) Fundamentals of ecology, 3rd edn. W.B. Saunders Co., Philadelphia

Chapter 5
Eutrophication Assessment Methods

Abstract Eutrophication is one of the most common causes of water quality impairment of inland and marine waters. Its best-known manifestations are toxic cyanobacteria blooms in lakes and waterways and proliferations of green macro algae in coastal areas. The term eutrophication is used by both the scientific community and public policymakers, and therefore has a myriad of definitions. The introduction by the public authorities of regulations to limit eutrophication is a source of tension and debate on the activities identified as contributing or having contributed decisively to these phenomena. Debates on the identification of the driving factors and risk levels of eutrophication, seeking to guide public policies, have led the ministries in charge of the environment and agriculture to ask for a joint scientific appraisal to be conducted on the subject. Four French research institutes were mandated to produce a critical scientific analysis on the latest knowledge of the causes, mechanisms, consequences and predictability of eutrophication phenomena. This paper provides the methodology and the main findings of this two years exercise involving 40 scientific experts.

Keywords Eutrophication · Nitrogen · Phosphorus · Algae bloom · Land-sea continuum · Diffuse pollution

Currently, there is no single opinion on the selection of parameters used to determine water trophic status, and there is no universal classification of different water trophic states. The first scientist who developed the concept of water trophic status in 1919 was the Swedish limnologist Einar Naumann from the University of Lund. Naumann based his original concept of classifying trophic states on the level of phytoplankton productivity and linked water trophic levels with physical and chemical factors influencing the development of phytoplankton in lakes (Table 5.1). These factors include temperature, light, calcium content, humic compounds, nitrogen, phosphorus, iron, pH, oxygen and carbon dioxide. He further determined the range of values of individual factors for each trophic type of water. Naumann emphasized that the basis for classification is productivity, but not the factors determining this productivity [1].

E. Neverova-Dziopak and Z. Kowalewski, *Eutrophication: A Global Environmental Problem*, Springer Water, https://doi.org/10.1007/978-3-031-83926-9_5

Table 5.1 Trophic types of lakes according to Naumann [2]

Trophic type	Water characteristics
Oligotrophic	Low productivity, low nitrogen and phosphorus content
Eutrophic	High productivity, high nitrogen and phosphorus content
Acidotrophic	Low productivity, pH value <5.5
Alkalitrophic	Low productivity, high calcium content
Agrillotrophic	Low productivity, high water turbidity
Siderotrophic	Low productivity, high iron content
Dystrophic	Low productivity, high content of humic compounds

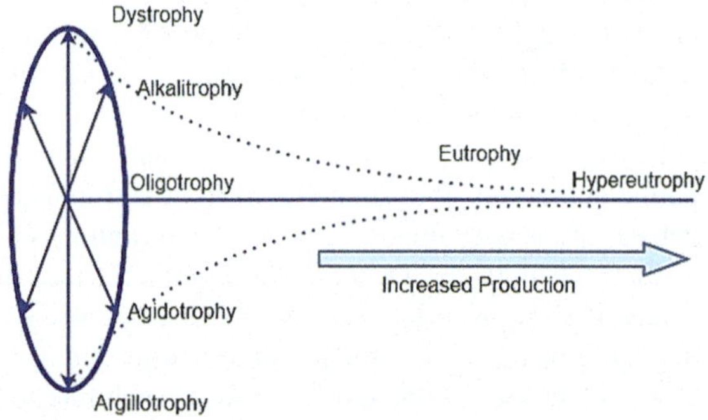

Fig. 5.1 Water typology according to Naumann [1]

Further limnological research allowed to state, that waters characterized by low productivity, as well as oligotrophic waters, differ significantly in terms of their biological structure, which allows to conclude, that the degree of productivity of acidotrophic, alkalitrophic and agidotrophic waters is determined not only by the content of biogenic substances, but by other substances and factors (Fig. 5.1).

Thienemann [3] assumed, that the typology of lakes should be developed taking into account climatic zones and temperature conditions. Further research allowed the selection of lakes with very different characteristics, therefore the typology of lakes and determining their trophic type became an extremely complicated task. Table 5.2 presents the characteristics of the basic trophic types of water. Due to the very different conditions and features of the aquatic ecosystems functioning, in addition to the basic types of water trophic states—dystrophic, oligotrophic and eutrophic, transitional states have also been determined—ultra-oligotrophic (between dystrophic and oligotrophic) and mesotrophic (between oligotrophic and eutrophic).

Integrating the Naumann-Thienemann theory, it can be concluded that the trophic state cannot be determined directly. For this purpose, various indicators are used that reflect or correlate with the "real" trophic state of waters. So far, various eutrophication indicators have been developed as a basis for assessing the trophic level of surface waters.

Table 5.2 Characteristics of water with different trophic status [4, 5]

Water characteristics	Dystrofic	Oligotrofic	Eutrofic
Transparency	Low	High	Low
Color	Darky-ellow/brown	Shappire/blue/green	Yellow-green
Suspensed solids content	High	Low	High
pH value	pH < 7.0	pH ≤ 7.0	pH ≥ 7.0
Saturation with oxygen	Low, oxygen profile similar like in eutrophic water	High, even in the hypolimnion it remains about 50–60%	Small, oxygen deficits in the hypolimnion
Biogenic matter content	Low	Moderate	High
Humic substances content	High	Low	Low/moderate
Calcium content	Low	Low	Moderate/high
Bottom sediments	Rich in allochthonous humic compounds	With a low content of organic substances	Rich in autochthonous organic ingredients, resembling putrefactive sludge
Coastal vegetation	Poor	Scanty	Rich
Plant plankton	Quantitatively poor, no blooms (or rare)	Quantitatively poor, rare blooms	Quantitatively abundant, blooms are frequent
Zooplankton	Scattered	Quantitatively poor	Quantitatively rich
Bottom fauna	Poor in species and quantity (10–20 individuals per 1 m^2 of bottom)	Various in species, not very rich in quantity (1000 individuals per 1 m^2 of bottom)	Undiversified in terms of species, rich in quantity (2000–10,000 individuals per 1 m^2 of bottom)
Evolution of reservoirs	They transform into raised bogs	They become eutrophic	They transform into ponds, meadow swamps, low peat bogs

5.1 Traditional Eutrophication Indicators

Traditionally, the trophic state of surface waters is assessed on the basis of a large number of hydrobiological characteristics which include, among others: biomass abundance, species composition of aquatic vegetation, photosynthesis rate, chlorophyll concentration, algal biomass, zoobenthos as well as the visibility of the Secchi disk, phosphorus and nitrogen concentrations, oxygen conditions, etc. Abovementioned eutrophication indicators are widely used in practice and are described in detail in numerous literature sources [6–10].

The limit values of the above-mentioned indicators given by different authors for individual trophic levels vary significantly, which is primarily due to the diversity of the geographical areas they studied and due to the fact, that the course of the eutrophication process in various aquatic ecosystems depends on their morphometric, hydrological, hydrobiological and climatic conditions and the type of catchment area development. The scale of discrepancies in the numerical values of basic trophic indicators is presented in Table 5.3.

However, the above-mentioned indicators are not suitable for direct use for various engineering analyses and applied tasks, including mathematical models used to practically solve problems of water resources protection and management. Most traditional hydrobiological characteristics reflect mainly the production side of these processes, but the state of the biotic balance, as already mentioned, also depends on the course of the decomposition processes of organic substances.

The limit values of indicators developed for individual trophic states are extremely difficult to determine. It largely depends on the type of lake or stream, their properties, location and climatic conditions. The authors' subjective preferences are often important, which makes it impossible to develop uniform and unambiguous evaluation systems.

An alternative to the traditionally used eutrophication indicators are the so-called aggregated indicators—trophic state indexes. The need to develop aggregated

Table 5.3 Range of values of selected trophic indicators according to literature data

Indcators	Ultra oligotrophic	Oligo trophic	Meso trophic	Eutrophic	Hiper trophic
Nitrogen, mg/dm^3	<0.2	0.2–0.4	0.3–0.65	0.5–1.5	>1.5
Phosphorus, mg/dm^3	<0.005	0.005–0.04	0.01–0.06	0.025–0.1	>0.1
Transparency, m	6–12	8–64	1.5–8.0	0.7–2.0	>0.7
Primary production, mg C/m^2/day	0–50	50–300	100–1000	250–1500	–
Annual phytoplankton production, g C/m^2	0–10	10–30	30–300	0–300	200–4000
Chlorophyll *a*, mg/dm^3	0–1	1–40	2–60	7–1000	>1000
Biomass of phytoplankton, g/m^3	<0.2	0.2–0.5	0.5–30	0.8–10	>10
Bacteria number, mln cells/dm^3	–	<0.5	0.5–1.5	2–4	–

eutrophication indicators is determined by the need to obtain the necessary information in order to use it for operational and effective management of water resources. They are based on the calculation of numerical values of indexes, elaborated on the base of correlations established by their authors between the basic eutrophication factors. Usually these factors are: chlorophyll-a, nitrogen and phosphorus. The values of indexes are developed for each trophic level, which allows for evaluation of specific trophic state. The purpose of using such indices is a more reliable assessment of water trophic state (which is not always possible when assessed based on a set of different indicators), adaptation to the specificity of separate aquatic ecosystems and the possibility of quick and easy water monitoring in accordance with legislative requirements.

TSI Index

One of the first of such aggregated indexes was the TSI (Trophic State Index) Carlson index [11, 12], sometimes called CTSI [13]. TSI also is a base for a large group of other eutrophication indexes because of its possibility to modifications. In main version TSI is based on three numerical values. But later other scientists add the categories of trophic status, new parameters and targeted to average the value of basic TSI. It is also a metric for measuring eutrophication potential [14]. Original form of Carlson equation elaborated in 1977, was based on following parameters: Secchi disc (SD), Chl-a and TP (Eqs. 5.1–5.3). The results was presented as numerical values of three values without calculating final mean value, and without assessing trophic status.

$$TSI(SD) = 10\left(6 - \frac{\ln SD}{ln2}\right) \tag{5.1}$$

$$TSI(CHl) = 10\left(6 - \frac{2.04 - 0.68 \ln Chl}{\ln 2}\right) \tag{5.2}$$

$$TSI(TP) = 10\left(6 - \frac{ln\frac{48}{TP}}{\ln 2}\right) \tag{5.3}$$

So called "mean version" of TSI (CTSI) [13], contains one single value as a mean of rest parameters (Eq. 5.4). But Carlson found this modification to be illogical and unnecessary [15].

$$CTSI = \frac{TSI(SD) + TSI(Chla) + TSI(TP)}{3} \tag{5.4}$$

As was mentioned before, Carlson's TSI allowed for modifications and new parameters to be introduced which enabled its adaptations to other conditions and water types, not only for lakes. Such variant of modification with total nitrogen added as a new fourth parameter of Eq. (5.4) was created by Kratzer and Brezonik (Eq. 5.5), and was used for trophic state assessment of Florida lakes [16, 17].

$$TSI(TN) = 54.45 + 14.43 * \ln(\mathrm{TN}) \tag{5.5}$$

The Burns TLI index (Trophic Lake Index) is considered as an extension of TSI idea. It was created for New Zealand lakes, based on four parameters (Eqs. 5.6–5.9) and final equation with mean value of index (Eq. 5.10). Values for equations are based on annual means [18].

$$TLc = 2.22 + 2.54\log(Chla) \tag{5.6}$$

$$TLs = 5.10 + 2.60\log\left(\frac{1}{SD} - \frac{1}{40}\right) \tag{5.7}$$

$$TLc = 0.218 + 2.92\log(TP) \tag{5.8}$$

$$TLn = -3.61 + 3.01\log(TN) \tag{5.9}$$

$$TLI = \frac{1}{4} * (TLc + TLs + TLp + TLn) \tag{5.10}$$

All multipliers in TSI equations are based on linear regressions, so they could be easily recalculated and adjusted to local conditions. Table 5.4 presents some local variations of Carlson TSI usage.

TRIX Index

Another significant and well know system of trophic indexes are the set based on TRIX index authored by Richard Vollenweider. TRIX was created in the 80s for Mediterranean Sea. It is, similarly to Carlson TSI, based on two groups of parameters: nutritional factors (minN—mineral forms of nitrogen and TP—total phosphorus) and productivity factors (Chla—chlorophyll-a and %DO—percent of oxygen saturation). It also contains two scaling factors: k = 1.5 and m = 1.2 for scaling TRIX values between 0 and 10 (Eq. 5.11) [36].

$$TRIX = \frac{(Log10[Chla * (100 - \%DO) * minN * TP + 1.5)}{1.2} \tag{5.11}$$

Sometimes the Eq. (5.11) can be presented as Eqs. (5.12) and (5.13):

$$TRIX = \frac{10}{12}\left[(logM_{chl} + 0.5) + (logM_{DO} + 1) + (logM_{DIN} - 0.5) + (logM_{TP} + 0.5)\right] \tag{5.12}$$

$$TRIX = \frac{k}{n}\sum_{i=1}^{n}\frac{(\log Mi - \log Li)}{(\log Ui - logLi)} \tag{5.13}$$

Table 5.4 Locations with usage of Carlson index

Places	Source
30 American lakes	[11]
Mexican lakes	[19]
Laguna Bays Philippines	[20]
Chinese lakes	[21]
Tropical rivers	[22]
China rivers	[23]
Baltic Sea border lakes	[24]
Lake Victoria	[25]
Estuaries in India	[26]
Tropical lakes in Indonesia	[27]
Salty lagoons in Morocco	[28]
Subtropical reservoirs in Brazil	[29]
Small ponds in Brazil	[30]
Crass lakes in China	[31]
Dam lakes	[32]
Egyptian lakes	[33]
Japanese lakes	[34]
Indian rivers	[35]

The Eqs. (5.12) and (5.13) are the versions of TRIX formulated in 1992 [37, 38].

In opposite to Carlson TSI index, TRIX index allows to determine trophic status of water body in terms of the classification adopted in limnology on the base on the calculated index value (Table 5.5) [39]. The values of TRIX index are presented in Table 5.5.

Similar to Carlson TSI, TRIX could be modified and adjust to local conditions. For example TRIX containing total nitrogen (TN) and phosphates (P-PO4) (not mineral nitrogen and total phosphorus) was modified and used for Greek Seas trophic state. The modified set of TRIX variants was also used for Caspian Sea trophic state assessment [40, 41]. Selected water bodies where TRIX was used is presented in Table 5.6.

Table 5.5 TRIX values

Water type	TRIX scale
Ultraoligotrophic	2–4
Oligotrophic	4–5
Mesotrophic	5–6
Eutrophic	6–8
Dystrophic	8>

Table 5.6 Location with usage of TRIX

Places	Sources
Adriatic Sea	[36]
Spain lagoons Llobregat's river delta	[42]
Mangrove estuaries Amazon river	[43]
Bahia lagoon Mexico	[44]
Rio de Plata estuary	[45]
South Atlantic coastal lagoon	[46]
Andean lakes	[47]
Tropical microtidal estuary	[48]

Another famous index of eutrophication is EI (Eutrophication Index) authored by Karydis [49]. EI was created in 80s and developed for Aegean Sea waters. EI index is based on TRIX idea adopted on the base of Principle Component Analysis (PCA) of data for Greek seas. In Eq. 5.14 C stands for concentrations and *a, b, c, d, e* are the coefficients derived from PCA analysis.

$$EI = aC_{\mathrm{PO_4}} + bC_{\mathrm{NO_3}} + cC_{\mathrm{NO_2}} + dC_{\mathrm{NH)3}} + eC_{\mathrm{chl\text{-}a}} \quad (5.14)$$

The index is not as popular as TRIX but it was mainly used in Mediterranean Sea [49, 50].

There is another group of indexes called EI, but they were created in China and have different construction. Chinses EI based on three parameters: COD, inorganic nitrogen (DIN) and inorganic phosphorus (DIP) (Eq. 5.15). EI from China is also available in local variants, with changes in parameters and value of coefficients [51–54].

$$EI = \frac{C_{COD} C_{DIN} C_{DIP}}{4500} 10^6 \quad (5.15)$$

Balance Based Indexes

Another methodology to assess the trophic state is based on the integral ITS (Index of Trophic State), which was developed for the purpose of monitoring the trophy of fresh waters in the Neva Bay of the Baltic Sea in 1995. The possibility of using ITS for different water types was verified when assessing the trophic state of all 26 types of 400 rivers in Poland, 7 Polish dam reservoirs, small rivers in Estonia, Caspian Sea, Saaler Boden in Germany and other aquatic systems [4, 55, 56].

ITS as one of the only few assessment methods based on the balance approach to eutrophication process assessment. The theoretical assumption of the ITS index is a change in the balance of primary organic matter production and decomposition processes (biotic balance) during eutrophication. Changes in the biotic balance are accompanied by changes in gas balance in waters and, consequently, in the quantitative ratios of oxygen and carbon dioxide concentrations. When the rate of decomposition of an organic substance exceeds the rate of its production, the concentration of carbon dioxide increases and oxygen concentration decreases; when the decomposition rate is slower than the production rate, the concentration of carbon dioxide decreases and oxygen concentration increases. This means that the change in the quantitative ratio of the content of these two gases in water reflects changes in biotic balance. In an aquatic environment, the oxygen content can be expressed by the oxygen saturation of the water, while the CO_2 content can be expressed by the pH. Changes in CO_2 concentrations leads to changes in pH, which results from carbonate balance, i.e. the state of a specific proportion between bicarbonate and bicarbonate ions, which is presented by the Eq. (5.16).

$$CO_2 + H_2 \leftrightarrow H_2CO_3 \leftrightarrow H^+ + HCO_3^- \leftrightarrow H^+ + CO_3^{2-} \quad (5.16)$$

In surface waters with low salinity in temperate climate zones, the pH value is usually between 6.0 and 8.5. Under such conditions, all forms of the carbon system may occur simultaneously. Carbon dioxide dissolved in water occurs mainly in the form of CO_2, only about 1% is in the form of carbonic acid H_2CO_3. Changes of CO_2 content in surface waters lead to a shift in the carbonate balance and the change in the concentration of hydrogen ions, and consequently—in the pH value of the water. When there is an excess of carbon dioxide, carbonates dissolve, and when there is a shortage of CO_2, they are deposited. When the speed of organic substance production processes is higher than the speed of its decomposition processes, CO_2 concentrations decrease and the pH value increases. With a deficiency of nutrients and low production, CO_2 concentrations increase and the pH value decreases [4].

The above considerations show that the trophic state can be described by the balance of CO_2 and O_2 concentrations, and these accordingly—by the pH value and the degree of water saturation with oxygen. The balance of pH values and water oxygen saturation is therefore the function of the biotic balance, i.e. the balance of production and decomposition processes. The relationship between pH and percentage saturation of water with oxygen for individual trophic states can be described by the established linear relationships (Fig. 5.2).

With the same oxygen saturation of the water, the higher the pH, the higher the trophic state. The pH value at normal 100% water saturation was chosen as the trophic criterion. The ITS index is calculated according to formula (5.17).

$$ITS = pH + a(100 - \%DO) \quad (5.17)$$

where: pH—mean pH value, %DO—mean oxygen saturation value, a—is a slope coefficient form pH-%DO linear regression.

The developed ITS values for waters of different trophic status are presented in the Table 5.7.

Similar approach based on mutual pH and oxygen saturation patterns was proposed much later by O'Boyle et al. [59]. The method was created and used for Irish inland lakes. Index equation is based on ranges of pH and water saturation with oxygen (Eq. 5.18).

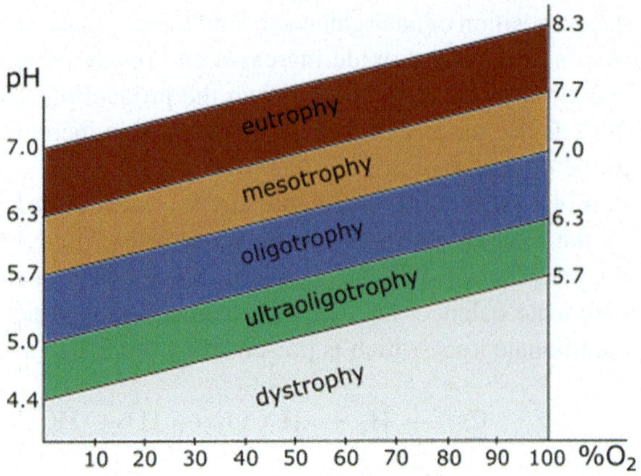

Fig. 5.2 ITS trophic states values [57]

Table 5.7 ITS trophic values [58]

Trophic status	ITS values
Dystrophy	<5.7–6.0
Ultraoligotrophy	6.0–6.6
Oligotrophy	6.7–73
Mezotrophy	7.4–8.0
Eutrophy	>8.0

$$O'Boyle\ Index = \frac{(range\,pH\,/\,\max\ range\,pH * 0.01) + (range\,DO/\max\ \mathrm{range\ DO} * 0.01)}{2} \tag{5.18}$$

O'Boyle Index values and their interpretation are presented in Table 5.8.

Table 5.8 Provisional scheme relating index values to different trophic [59]

Indicative trophic state	O'Boyle index value
Eutrophic	>75–100
Potentially eutrophic	>60 < 75
Intermediate	>40 < 60
Unpolluted	0 < 40

5.2 Hybrid Assessment

The third group of methods for assessing the trophic state are the assessment tools that can be called hybrid. They consist of groups of established standards, norms and interconnections implemented using simple models. They can be treated rather as "programs of control and assessment", not build up versions of integrated methods.

HEAT method

HEAT is an acronym for the HELCOM Eutrophication Assessment Tool [60] developed for assessment of eutrophication status of the water areas embraced by Helsinki Convention, i.e. the Baltic Sea. HEAT classification defines the water areas as either "areas affected by eutrophication" (having a moderate, poor or bad status) or 'areas not affected by eutrophication' (having a high or good status). HEAT is multi-indicator based tool using the "one out—all out" principle assumed by Water Framework Directive, which means that the overall classification of an assessed area is based on the most sensitive (weak) quality element and there is a lack of typical ecological status assessment.

HEAT is based on existing indicators, which for this purpose have been grouped as follows: (1) physical–chemical features (PC), (2) phytoplankton (PP), (3) submerged aquatic vegetation (SAV) and (4) benthic invertebrate communities (BIC). Groups 1 and 2 (PC and PP) are considered to be 'primary signals' of eutrophication, while groups 3 and 4 (SAV and BIC) are considered to be "secondary signals".

For each individual indicator, an interim classification is made. The classification system has five classes: high and good corresponding to "areas not affected by eutrophication" and moderate, poor and bad, which correspond to "areas affected by eutrophication". All indicators are divided in three groups with reference conditions (RefCon), acceptable deviations (AcDev) and actual status (AcStat). RefCon, which are "a description of the biological quality elements that exist (or would exist) at high status, that is, with no (or very minor) disturbance from human activities". They are used to quantify the degree of disturbance observed in the environment. Furthermore, they should represent part of nature's continuum and must reflect variability [61]. Idea of HEAT EQR (Ecological Quiality Ratio) is shown on Fig. 5.3.

ASSETS tool

The Assessment of Estuarine Trophic Status (ASSETS) assessment method is based on Pressure-State-Response model, that has been used most recently to complete the update of the National Estuarine Eutrophication Assessment (NEEA) [62]. The basic idea of ASSETS assessment is present on Fig. 5.4.

WQI system of assessment

WQI (Water Quality Index) is a large group of multi indexes with few specific methodologies, some of them refer to eutrophication process. The main idea of WQI is a relation between chosen parameters with adding a specific weights to them. For example, NSF WQI (National Sanitation Foundation Water Quality Index), includes

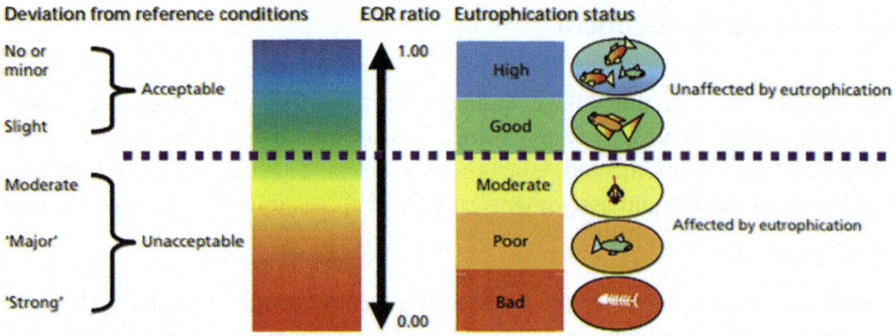

Fig. 5.3 HELCOM EQR concept ([60] CY BB)

Fig. 5.4 Basic idea of ASSETS

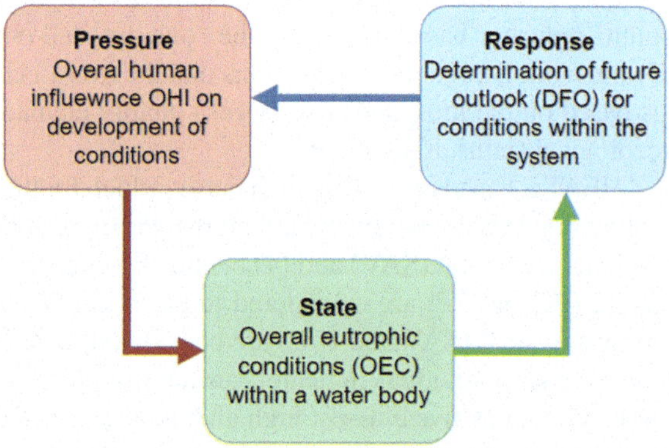

7 physico-chemical parameters. This method uses the following formula, where the parameters are valued on a scale of 0 to 100, and the ratio of the number of substandard results (to the total number of results (qi) and the weight for each of them (wi): i.e., DO = 0.24, BOD_5 = 0.20, COD = 0.15, TP = 0.17, EC = 0.11, NO_3^-N = 0.09, and pH = 0.04, is calculated as square weighted mean.

UWQI (Universal Water Quality Index) is calculated from the results for 5 physicochemical parameters (BOD_5, NO_3^-, DO, TP, and pH) from the considered multi-year period with 90th or 10th percentile.

OWQI (Oregon Water Quality Index) is calculating by taking medians for 6 parameters (in this case: DO, BOD_5, $NH_4 + NO_3^-N$, TP, pH, and temperature of water) [63].

TWQI (Transitional Water Quality Index) is a multi-metric index that integrates main causal factors (N and P concentrations), key biological elements (phytoplankton as chlorophyll-a, benthic phanerogams and macroalgal cover) and indicators of eutrophication effects (dissolved oxygen saturation). Measured parameters are modificated using non linear quality function for calculating quality values, then results are weighted, seasonality is included [63].

The next method created based on TWQI is TWEAM (Transitional Water Eutrophication Assessment). TWEAM is a multi index based on the ecological index MaQI (Macrophyte Quality Index), the trophic index TWQI (Transitional Water Quality Index), and physicochemical quality elements (dissolved inorganic nitrogen (DIN) and orthophosphates ($P\text{-}PO_4$). It contains three phases of assessment contacted with expert assessments. In phase two there is a directly trophic status described as: non-eutrophic, mesotrophic or eutrophic. A final results is a holistic assessment referring to the following ecological states: bod, poor, moderate, good, high. Both method were created in Italy, for Mediterranean Sea costal waters [64].

OSPAR procedure

Oslo and Paris Convention Common Procedure (OSPAR COMPP) assumes two stages of assessment: initial screening of areas and comprehensive assessment. Screening process qualifies water areas in following categories: (1) non problem, (2) potential problem and (3) problem areas. The indicators for these categories include: (1) nutrient enrichment (nutrient inputs, DIN and DIP concentrations in winter period); (2) direct effects of nutrient enrichment (Chl-a concentration, elevated levels of toxic phytoplankton indicator species and macrophytes and shift from long-lived to short-lived nuisance macrophytes species); (3) indirect effects of nutrient enrichment (oxygen deficiency, fish kills and long-term area-specific changes in zoobenthos biomass and fish, area-specific elevated levels of organic carbon/organic matter in relation to oxygen deficiency; (4) other possible effects of nutrient enrichment, includes algal toxins and transboundary transport [65]

IFREMER Methods

The French National Institute for Ocean Science And Technology (IFREMER) method uses the 90th percentile of annual or seasonal Chl-a data which is compared to a fixed scale reference conditions that is determined from studies (BY OECD, EPANCA, IFREMER). It was developed for French Mediterranean lagoons [65, 66].

EPA NCCA System

The National Coastal Condition Assessment (NCCA) is a national coastal monitoring program with rigorous quality assurance protocols and standardized sampling procedures designed to produce national and regional assessment of coastal condition. This system of water quality assessment is based on the measurements of nutrients, chlorophyll-a, dissolved oxygen and water clarity to characterize the probability of water body experiencing one of three stages of eutrophication: pre-algal bloom, bloom or post-bloom. Using regionally relevant benchmarks, EPA assigned each individual parameter a rating of good, fair or poor [67].

5.3 The Requirements for Eutrophication Indicators

Environmental indicators are the main tool for assessing the state of water ecosystems and setting the directions of water policy. Appropriately selected indicators based on extensive resources of data accumulated from long-term environmental observations and monitoring will reflect basic trends and allow establishing the cause-and-effect relationships with changes in the aquatic environment state and ecological conditions. They also make it possible to track and assess the effectiveness of the implementation of ecological policy, make changes and corrections to it and adopt to the current situation, define priority goals and tasks, as well as develop target norms and standards for water quality and status. Such indicators should reflect the functions of water ecosystems and different types of water use, characterize the quality of water, the impact on their individual elements, the scale of pollution and reflect the effectiveness of undertaken protective actions.

Anthropogenic eutrophication indicators are the features reflecting various aspects of disturbances in aquatic ecosystems under the influence of anthropogenic factors. These indicators should primarily reflect those type of disturbances that are related to changes in the abundance of water in nutrients, except the environmental changes, occurring as a result of seasonal phenomena or short-term climate cycles. These indicators should be sensitive to changes in the content of nutrients and universal for various types of water bodies in order to compare the results of assessment.

A very important practical feature of such indicators should be a simple way of measuring, interpreting and assessing, and they should be useful for various practical purposes, i.e. for monitoring, forecasting, mathematical modeling, assessing the effectiveness of protection measures and making decisions regarding sustainable use of water resources.

The basic requirements for environmental indicators were formulated by the Working Group for Monitoring and Evaluation of the United Nations Economic Commission for Europe (UNECE), UNESCO Intergovernmental Oceanographic Commission and by Publications Office of the European Union [68–70]. The mentioned requirements are presented in Table 5.9.

Table 5.9 Requirements for environmental indicators

UNECE	UNESCO	EU Publications Office
• Indicators should be developed in a form accessible to all users • Indicators should reflect phenomena or processes without unnecessary details • Access to the necessary scope of data should be ensured to provide the credibility of the indicators	• Indicator should contain information about the phenomenon/process and be based on the minimum possible number of measurements, but sufficient to present the essence of the phenomenon or process • Indicator should be easy to interpret and understandable not only for scientists, but also for institutions decisions-makers in the field of environmental protection and management and for communities	• Indicators should reflect changes in aquatic ecosystems caused by anthropogenic factors • Reflect the basic functions of aquatic ecosystems • Be based on a limited amount of available data and the cost of evaluating it should be moderate • Be suitable for predicting changes in the state of ecosystems • Reference values should be developed for indicators in order to enable interpretation of possible changes or threats by various users of these indicators • They should be aggregated, integrating information in time and space • They should be theoretically and technically justified

References

1. Naumann E (1929) The scope and chief problems of regional limnology. Internationale Revue Der Gesamten Hydrobiologie Und Hydrographie 22(1):423–444. https://doi.org/10.1002/iroh.19290220128
2. Naumann E (1933) Grundzüge der regionalen Limnologie. Nature 131(3303):223–224. https://doi.org/10.1038/131223a0
3. Thienemann A (1921) Seetypen. Die Naturwissenschaften 9(18):343–346. https://doi.org/10.1007/BF01487893
4. Neverova-Dziopak E (2010) Podstawy zarządzania procesem eutrofizacji antropogenicznej. Wydawnictwa AGH
5. Pliński M (1992) Hydrobiologia ogólna. Wydawnictwo Uniwersytetu Gdańskiego
6. Bricker SB, Ferreira JG, Simas T (2003) An integrated methodology for assessment of estuarine trophic status. Ecol Model 169(1):39–60. https://doi.org/10.1016/S0304-3800(03)00199-6
7. Cheng K, Lei T (2001) Reservoir trophic state evaluation using Landsat TM images. JAWRA J Am Water Resour Assoc 37(5):1321–1334. https://doi.org/10.1111/j.1752-1688.2001.tb03642.x
8. Dodds WK, Jones JR, Welch EB (1998) Suggested classification of stream trophic state: distributions of temperate stream types by chlorophyll, total nitrogen, and phosphorus. Water Res 32(5):1455–1462. https://doi.org/10.1016/S0043-1354(97)00370-9
9. Hering D, Johnson R, Kramm S, Schmutz S, Szoszkiewicz K, Verdonschot P (2006) Assessment of European streams with diatoms, macrophytes, macroinvertebrates and fish: a comparative metric-based analysis of organism response to stress. Freshw Biol 51(9):1757–1785. https://doi.org/10.1111/j.1365-2427.2006.01610.x

10. Yang X, Wu X, Hao H, He Z (2008) Mechanisms and assessment of water eutrophication. J Zhejiang Univ Sci B 9(3):197–209. https://doi.org/10.1631/jzus.B0710626
11. Carlson RE (1977) A trophic state index for lakes. Limnol Oceanogr 22(2):361–369. https://doi.org/10.4319/lo.1977.22.2.0361
12. Carlson RE (2007) Estimating trophic state. https://www.researchgate.net/publication/263042353
13. Lin JL, Karangan A, Huang YM, Kang SF (2022) Eutrophication factor analysis using Carlson trophic state index (CTSI) towards non-algal impact reservoirs in Taiwan. Sustain Environ Res 32(1). https://doi.org/10.1186/s42834-022-00134-x
14. Tsuzuki Y (2006) An attempt of modification Carlson's trophic state index (TSI) for brackish lakes in Japan. Laguna 13(6):89–98. www.scoups.com
15. Carlson R, Simpson J (1996) A coordinator's guide to volunteer lake monitoring methods. North American Lake Management Society
16. Brezonik PL (1984) Trophic state indices: rationale for multivariate approaches. Lake Reserv Manag 1(1):441–445. https://doi.org/10.1080/07438148409354553
17. Kratzer CR, Brezonik PL (1981) A Carlson-type trophic state index for nitrogen in Florida Lakes. JAWRA J Am Water Resour Assoc 17(4):713–715. https://doi.org/10.1111/j.1752-1688.1981.tb01282.x
18. Burns NM, Rutherford JC, Clayton JS (1999) A monitoring and classification system for New Zealand lakes and reservoirs. Lake Reserv Manag 15(4):255–271. https://doi.org/10.1080/07438149909354122
19. Membrillo-Abad A-S, Torres-Vera M-A, Alcocer J, Prol-Ledesma RM, Oseguera L, Ruiz-Armenta JR (2016) Trophic State Index estimation from remote sensing of lake Chapala, México. Revista Mexicana de Ciencias Geológicas 33:183–191
20. Escober EJ, Espino MP (2023) A new trophic state index for assessing eutrophication of Laguna de Bay, Philippines. Environ Adv 13:100410. https://doi.org/10.1016/j.envadv.2023.100410
21. Zhang D, Shi K, Wang W, Wang X, Zhang Y, Qin B, Zhu M, Dong B, Zhang Y (2024) An optical mechanism-based deep learning approach for deriving water trophic state of China's lakes from Landsat images. Water Res 252:121181. https://doi.org/10.1016/j.watres.2024.121181
22. da Costa Lobato T, Hauser-Davis RA, de Oliveira TF, Maciel MC, Tavares MRM, da Silveira AM, Saraiva ACF (2015) Categorization of the trophic status of a hydroelectric power plant reservoir in the Brazilian Amazon by statistical analyses and fuzzy approaches. Sci Total Environ 506–507:613–620. https://doi.org/10.1016/j.scitotenv.2014.11.032
23. Li X, Li F, Min X, Xie Y, Zhang Y (2023) Embracing eDNA and machine learning for taxonomy-free microorganisms biomonitoring to assess the river ecological status. Ecol Ind 155:110948. https://doi.org/10.1016/j.ecolind.2023.110948
24. Szymańska-Walkiewicz M, Matela M, Obolewski K (2024) Patterns of effects of land-use structure on lake water quality in coastal lake catchments of the southern Baltic Sea. Ecohydrol Hydrobiol 24(2):447–458. https://doi.org/10.1016/j.ecohyd.2023.07.004
25. Gidudu A, Letaru L, Kulabako RN (2021) Empirical modeling of chlorophyll a from MODIS satellite imagery for trophic status monitoring of Lake Victoria in East Africa. J Great Lakes Res 47(4):1209–1218. https://doi.org/10.1016/j.jglr.2021.05.005
26. Niveditha SK, Haridevi CK, Hardikar R, Ram A (2022) Phytoplankton assemblage and chlorophyll a along the salinity gradient in a hypoxic eutrophic tropical estuary-Ulhas Estuary, West Coast of India. Mar Pollut Bull 180:113719. https://doi.org/10.1016/j.marpolbul.2022.113719
27. Tanjung RHR, Indrayani E, Agamawan LPI, Hamuna B (2024) Water quality assessment to determine the trophic state and suitability of Lake Sentani (Indonesia) for various utilisation purposes. Water Cycle 5:99–108. https://doi.org/10.1016/j.watcyc.2024.02.006
28. Ngadi H, Layachi M, Azizi G, Baghour M, Esseffar S, Loukili H, Moumen A (2023) Evaluation of the water quality and the eutrophication risk in Ramsar site on Moroccan northern Mediterranean (Marchica lagoon): a multivariate statistical approach. Mar Pollut Bull 194:115373. https://doi.org/10.1016/j.marpolbul.2023.115373
29. Cunha DGF, do Calijuri MC, Lamparelli MC (2013) A trophic state index for tropical/subtropical reservoirs (TSItsr). Ecol Eng 60:126–134. https://doi.org/10.1016/j.ecoleng.2013.07.058

30. Wiegand MC, do Nascimento ATP, Costa AC, Lima Neto IE (2021) Trophic state changes of semi-arid reservoirs as a function of the hydro-climatic variability. J Arid Environ 184:104321. https://doi.org/10.1016/j.jaridenv.2020.104321
31. Ni M, Ge Q, Li S, Wang Z, Wu Y (2021) Trophic state index linked to partial pressure of aquatic carbon dioxide in a typical karst plateau lake. Ecol Ind 120:106912. https://doi.org/10.1016/j.ecolind.2020.106912
32. Talukdar S, Pal S, Chakraborty A, Mahato S (2020) Damming effects on trophic and habitat state of riparian wetlands and their spatial relationship. Ecol Ind 118:106757. https://doi.org/10.1016/j.ecolind.2020.106757
33. Zaghloul GY, Zaghloul AY, Hamed MA, El-Moselhy KM, El-Din HME (2023) Water quality assessment for Northern Egyptian lakes (Bardawil, Manzala, and Burullus) using NSF-WQI Index. Reg Stud Mar Sci 64:103010. https://doi.org/10.1016/j.rsma.2023.103010
34. Aizaki M, Otsuki A, Fukushima T, Hosomi M, Muraoka K (1981) Application of Carlson's trophic state index to Japanese lakes and relationships between the index and other parameters. SIL Proc 1922–2010 21(1):675–681. https://doi.org/10.1080/03680770.1980.11897067
35. Khatun R, Talukdar S, Pal S, Kundu S (2021) Measuring dam induced alteration in water richness and eco-hydrological deficit in flood plain wetland. J Environ Manag 285:112157. https://doi.org/10.1016/j.jenvman.2021.112157
36. Vollenweider RA, Giovanardi F, Montanari G, Rinaldi A (1998) Characterization of the trophic conditions of marine coastal waters with special reference to the NW Adriatic Sea: proposal for a trophic scale, turbidity and generalized water quality index. Environmetrics 9(3):329–357. https://doi.org/10.1002/(SICI)1099-095X(199805/06)9:3%3c329::AID-ENV308%3e3.0.CO;2-9
37. Fiori E, Zavatarelli M, Pinardi N, Mazziotti C, Ferrari CR (2016) Observed and simulated trophic index (TRIX) values for the Adriatic Sea basin. Nat Hazard 16(9):2043–2054. https://doi.org/10.5194/nhess-16-2043-2016
38. Vollenweider RA, Rinaldi A, Montanari G (1992) Eutrophication, structure and dynamics of a marine coastal system: results of ten-year monitoring along the Emilia-Romagna coast (Northwest Adriatic Sea). In: Marine coastal eutrophication. Elsevier, pp 63–106. https://doi.org/10.1016/B978-0-444-89990-3.50014-6
39. Pettine M, Casentini B, Fazi S, Giovanardi F, Pagnotta R (2007) A revisitation of TRIX for trophic status assessment in the light of the European water framework directive: application to Italian coastal waters. Mar Pollut Bull 54(9):1413–1426. https://doi.org/10.1016/j.marpolbul.2007.05.013
40. Nasrollahzadeh Saravi H, Pourang N, Foong SY, Makhlough A (2019) Eutrophication and trophic status using different indices: a study in the Iranian coastal waters of the Caspian Sea. Iranian J Fish Sci. https://doi.org/10.22092/ijfs.2018.117717
41. Primpas I, Karydis M (2011) Scaling the trophic index (TRIX) in oligotrophic marine environments. Environ Monit Assess 178(1–4):257–269. https://doi.org/10.1007/s10661-010-1687-x
42. Cañedo-Argüelles M, Rieradevall M, Farrés-Corell R, Newton A (2012) Annual characterisation of four Mediterranean coastal lagoons subjected to intense human activity. Estuar Coast Shelf Sci 114:59–69. https://doi.org/10.1016/j.ecss.2011.07.017
43. Pereira LCC, Jiménez JA, Pineda MG, da Costa ÁKR, Sousa N, de Oliveira ARG, da Costa RM (2023) Assessment of trophic status in Amazonian estuaries: a reinterpretation of TRIX values in mangrove estuaries. Ocean Coast Manag 244:106805. https://doi.org/10.1016/j.ocecoaman.2023.106805
44. Cervantes-Duarte R, del Jimenez-Quiroz MC, Funes-Rodriguez R, Hernandez-Trujillo S, Gonzalez-Armas R, Anaya-Godinez E (2021) Interannual variability in the trophic status and water quality of Bahía Magdalena, Mexico, during the 2015–2018 period: TRIX. Reg Stud Mar Sci 42:101638. https://doi.org/10.1016/j.rsma.2021.101638
45. Brugnoli E, Muniz P, Venturini N, Brena B, Rodríguez A, García-Rodríguez F (2019) Assessing multimetric trophic state variability during an ENSO event in a large estuary (Río de la Plata, South America). Reg Stud Mar Sci 28:100565. https://doi.org/10.1016/j.rsma.2019.100565

46. Tudurí A, Becoña LP, Venturini N, Rodríguez-Gallego L, García-Rodríguez F, González L, Lescano C, Costa S, del Puerto L, Bergamino L (2021) Trophic assessment in South American Atlantic coastal lagoons: linking water, sediment and diatom indicators. Mar Pollut Bull 165:112119. https://doi.org/10.1016/j.marpolbul.2021.112119
47. García-Avila F, Loja-Suco P, Siguenza-Jeton C, Jiménez-Ordoñez M, Valdiviezo-Gonzales L, Cabello-Torres R, Aviles-Añazco A (2023) Evaluation of the water quality of a high Andean lake using different quantitative approaches. Ecol Ind 154:110924. https://doi.org/10.1016/j.ecolind.2023.110924
48. Gogoi P, Das SK, Jana C, Das BK, Saha A, Ramteke K, Jaiswar AK, Samanta S, Roshith CM (2024) Assessing the trophic status of a tropical microtidal estuary applying TRIX and random forest—a combined approach. Mar Pollut Bull 200:116126. https://doi.org/10.1016/j.marpolbul.2024.116126
49. Karydis M, Ignatiades L, Moschopoulou N (1983) An index associated with nutrient eutrophication in the marine environment. Estuar Coast Shelf Sci 16(3):339–344. https://doi.org/10.1016/0272-7714(83)90151-8
50. Primpas I, Tsirtsis G, Karydis M, Kokkoris GD (2010) Principal component analysis: development of a multivariate index for assessing eutrophication according to the European water framework directive. Ecol Ind 10(2):178–183. https://doi.org/10.1016/j.ecolind.2009.04.007
51. Dou Y, Zhou WL (2022) Community characteristics of phytoplankton and eutrophication assessment in Tianjin section, downstream of Haihe River Basin. J Freshw Ecol 37(1):525–542. https://doi.org/10.1080/02705060.2022.2118179
52. Jiang Q, He J, Wu J, Hu X, Ye G, Christakos G (2019) Assessing the severe eutrophication status and spatial trend in the coastal waters of Zhejiang province (China). Limnol Oceanogr 64(1):3–17. https://doi.org/10.1002/lno.11013
53. Wang F, Zhang R, Kang P, Zhao H (2018) An evaluation of the ecological security of the Dongting Lake, China. Desalin Water Treat 110:283–297. https://doi.org/10.5004/dwt.2018.22328
54. Zhang W, Jin X, Liu D, Lang C, Shan B (2017) Temporal and spatial variation of nitrogen and phosphorus and eutrophication assessment for a typical arid river—Fuyang River in northern China. J Environ Sci (China) 55:41–48. https://doi.org/10.1016/j.jes.2016.07.004
55. Neverova-Dziopak E, Kowalewski Z (2014) New approach to trophic state assessment of running waters in Poland. Meteorol Hydrol Water Manag 1(1):15–22. https://doi.org/10.26491/mhwm/20551
56. Preisner M, Neverova-Dziopak E, Kowalewski Z (2020) Mitigation of eutrophication caused by wastewater discharge: a simulation-based approach. Ambio. https://doi.org/10.1007/s13280-020-01346-4
57. Neverova-Dziopak E, Kowalewski Z, Preisner M (2023) The universal trophic index: new methodological approach to eutrophication monitoring and control. Aquat Sci 85(1):6. https://doi.org/10.1007/s00027-022-00901-3
58. Neverova-Dziopak E (2010) Basics of managing the process of anthropogenic eutrophication. AGH
59. O'Boyle S, McDermott G, Noklegaard T, Wilkes R (2013) A simple index of trophic status in estuaries and coastal bays based on measurements of pH and dissolved oxygen. Estuaries Coasts 36(1):158–173. https://doi.org/10.1007/s12237-012-9553-4
60. HELCOM (2009) Eutrophication in the Baltic Sea
61. Andersen JH, Axe P, Backer H, Carstensen J, Claussen U, Fleming-Lehtinen V, Järvinen M, Kaartokallio H, Knuuttila S, Korpinen S, Kubiliute A, Laamanen M, Lysiak-Pastuszak E, Martin G, Murray C, Møhlenberg F, Nausch G, Norkko A, Villnäs A (2011) Getting the measure of eutrophication in the Baltic Sea: towards improved assessment principles and methods. Biogeochemistry 106(2):137–156. https://doi.org/10.1007/s10533-010-9508-4
62. Bricker S, Ferreira J (2007) ASSETS eutrophication assessment: method and application. In: Meeting with Venice water authority
63. Buta B, Wiatkowski M, Gruss Ł, Tomczyk P, Kasperek R (2023) Spatio-temporal evolution of eutrophication and water quality in the Turawa dam reservoir, Poland. Sci Rep 13(1):9880. https://doi.org/10.1038/s41598-023-36936-1

64. Bonometto A, Ponis E, Cacciatore F, Riccardi E, Pigozzi S, Parati P, Novello M, Ungaro N, Acquavita A, Manconi P, Sfriso A, Giordani G, Boscolo Brusà R (2022) A new multi-index method for the eutrophication assessment in transitional waters: large-scale implementation in Italian lagoons. Environments 9(4):41. https://doi.org/10.3390/environments9040041
65. Brehmer P, Do Chi T, Laugier T, Galgani F, Laloë F, Darnaude AM, Fiandrino A, Mouillot D (2011) Field investigations and multi-indicators for shallow water lagoon management: perspective for societal benefit. Aquat Conserv Mar Freshw Ecosyst 21(7):728–742. https://doi.org/10.1002/aqc.1231
66. Le Moal M, Gascuel-Odoux C, Ménesguen A, Souchon Y, Étrillard C, Levain A, Moatar F, Pannard A, Souchu P, Lefebvre A, Pinay G (2019) Eutrophication: A new wine in an old bottle?, Science of The Total Environment 651:1–11. ISSN 0048-9697, https://doi.org/10.1016/j.scitotenv.2018.09.139. https://www.sciencedirect.com/science/article/pii/S0048969718335836
67. US EPA (2021) A collaborative survey of the nation's estuaries and great lakes nearshore waters. National coastal condition assessment
68. Belfiore S, Balgos M, Mclean B, Galofré J, Blaydes M, Tesch D (2003) A reference guide on the use of indicators for integrated coastal management
69. EC (2000) Towards environmental pressure indicators for the EU
70. UNECE (2008) Ninth session of the working group on environmental monitoring and assessment

Chapter 6
Analysis of Existing Approaches in Eutrophication Management

Abstract This chapter presents an in-depth critical analysis of current approaches and strategies for managing eutrophication worldwide. The authors try to analyse the reasons for the failure to combat the globally intensifying phenomenon of eutrophication, despite the actions and efforts undertaken to mitigate this process and its negative effects. The result of the undertaken analysis are selected aspects of this phenomenon which, according to the authors, constitute a "weak link" in the strategy for process managing and considerations about them.

Keywords Monitoring and assessment · Limiting factors · Assimilation capacity · Ecological rating · Recipient oriented approach

6.1 Problems of Monitoring and Assessment

In order to make a decision on rational water management and protection against eutrophication it is necessary, above all, to have an objective information on the trophic status of water ecosystem. The condition of natural waters ought to be monitored in order to provide the competent authorities and water users with reliable information on the state of water bodies and water quality. The control of trophic state dynamics in aquatic ecosystems is a task of special ecological monitoring, which should be based on the principles other than regular monitoring of environmental pollution.

The existing system of surface water observation in European and other countries is necessary but insufficient for the purposes of trophic state monitoring. It needs not only some improvements and additional operations to eliminate the asynchrony of hydrological, chemical, bacteriological and hydrobiological observations and analyses, but also the introduction of automatic tracking and alert systems, the use of remote surveillance methods, etc.

Moreover, considering the balance approach to the essence of eutrophication processes, it can be stated, that the criterion, which in a most adequate way reflects the ecological trophic status of surface water is material, energetic or kinetic balance

E. Neverova-Dziopak and Z. Kowalewski, *Eutrophication: A Global Environmental Problem*, Springer Water, https://doi.org/10.1007/978-3-031-83926-9_6

of production and decomposition of organic matter, because the level of water trophy is conditioned by the final balance of these processes. At the same time, this criterion characterizes the ecological state of water ecosystem. This approach shows that both the state of biotic balance, as well as trophic state in this context are the synonyms [1].

As shows the state of knowledge analysis, an appropriate assessment of water trophic status, the scale of negative changes in aquatic ecosystem and water quality is difficult due to the lack of sufficient long-term results of hydrobiological parameters observations. In addition, the system of eutrophication monitoring is unsatisfactory: there is no uniform methodological approach to trophic status assessment, the lack of reliable and low-cost eutrophication indicators used.

Regarding the European legislation on the status of aquatic environment, it should be noted that many EU directives impose an obligation on member states to realize the monitoring of the parameters important for eutrophication assessment and appropriate recommendations on their limit values. The assessment of trophic state according to the recommendations of different EU Directives often gives contradictory results. This is mainly the effect of the difference in adopted assumptions and approaches to assessing this phenomenon and practical objectives of the assessment. The different set of suggested trophic indicators and different interpretation of their reference values also contributes to the discrepancy of assessment results (Table 6.1) [2].

None of these directives gives a unified definition of the process, nor a methodology for its evaluation. Except this, various indicators with different limit values are applied in different countries to assess the course of this phenomenon. In the interpretation of EU Directives, eutrophication is an undesirable phenomenon and does not require the precise definition of trophic state, but only the statement of the occurrence of adverse changes in water status. According to these documents the trophic state is defined in descriptive categories (Table 6.2).

It can be concluded that, according to the Water Framework Directive, the trophic status of water may correspond to moderate, weak or bad water status, depending on the scale of changes in the biological elements of the assessment.

The analysis of methodological problems related to the assessment of the degree of eutrophication showed, that neither the EU Directives nor regulations in the EU countries provide a uniform methodology for assessing this phenomenon. In EU countries, the approach developed by the OECD is most often used to assess the degree of eutrophication of surface waters. The assessment of the degree of eutrophication is included indirectly, as an element of the assessment of their ecological status based on selected physico-chemical and biological parameters [10].

The main element of eutrophication monitoring is the organization of a spatial and temporal observation network, on which the degree of the received information representativeness completely depends. The main goal of monitoring is to assess and predict the trophic state of water ecosystem [12].

Solving these problems requires an in-depth analysis of the obtained observational results based on theoretical concepts about the patterns of spatiotemporal changes in the observed parameters. Due to the lack of knowledge about the patterns of matter

Table 6.1 Requirements of EU Directives regarding surface waters eutrophication assessment and the scope of monitoring [3–10]

Directive	Requirements for the assessment of eutrophication	Minimum requirements for monitoring of eutrophication
Water framework directive 2000/60/EU	Considerate in the assessment of ecological status in case when the enrichment in biogenic substances effects the biological and physical and chemical quality elements	Phytoplankton (every 6 months), aquatic vegetation, macro invertebrates, fish (every 3 years) Hydromorphological elements (hydrology—continuous or monthly) other (every 6 years) Elements of physical and chemical quality (every 3 years)
Directive 91/271/EEC on urban wastewater treatment	Conducted in order to identify the areas sensitive to eutrophication (i.e. eutrophic waters or which may become eutrophic in the near future)	Control of the sensitive areas or identification of new ones every 4 years
Directive 91/676/EEC concerning the protection of waters against pollution caused by nitrates from agricultural sources	Conducted in order to identify areas particularly exposed to pollution by nitrogen compounds from agricultural sources and eutrophication	Monitoring of nitrate content in surface waters and assessment of trophic status (every 4 years)
Directive 2006/44/EC fish directive	Does not contain the specific requirements for the assessment of eutrophication, but gives the recommended values of phosphorus concentrations	Ammonia, nitrites, pH, total phosphorus, dissolved oxygen (once per month)
Directive 79/923/EEC on the quality required of shellfish waters	Does not specify the requirements for the assessment of eutrophication	pH, dissolved oxygen (once per month)
Directive 2016/11/EU on pollution caused by some dangerous substances discharged to the aquatic environment	No special requirements for eutrophication assessment but qualitative requirements for phosphorus and substances affecting oxygen conditions, especially ammonia and nitrite	No special requirements
Directive 92/43/EEC on the conservation of natural habitats and of wild fauna and flora	Assessment of eutrophication only if there is a threat to the habitat or species	No special requirements

and energy circulation during the process of eutrophication, even the analysis of observational results is a difficult task [13, 14]. The organization of an effective observation system in any case involves the solution of few interrelated problems answering four strategic questions (Table 6.3).

Table 6.2 Trophic state assessment according to EU Directives (on the base of [11])

Ecological status	Water framework directive	Wastewater directive	Nitrates directive	OSPAR
Very good	Almost undisturbed conditions	Non-eutrophic, no sensitive zone assessing required	Non-eutrophic, non-polluted, assessing of nitrate-sensitive zones is not required	No problem zone
Good	Minor changes in the composition and biomass of essential biological elements	Non-eutrophic, no sensitive zone assessing required	Non-eutrophic, non-polluted, assessing of nitrate-sensitive zones is not required	No problem zone
Moderate	Moderate changes in the composition and biomass of relevant biological elements	Eutrophic or likely to become eutrophic in the near future, marking of sensitive zones is required	Eutrophic or likely to become eutrophic in the near future, polluted, marking of nitrate sensitive zones is required	Problem zone
Weak	Significant changes in biological elements	Eutrophic, assessing of sensitive zones is required	Polluted, marking of nitrate sensitive zones is required	Problem zone
Bad	Major changes in biological elements	Eutrophic, assessing of sensitive zones is required	Polluted, assessing of nitrate sensitive zones is required	Problem zone

All of the above listed tasks of organizing the trophic state monitoring system are usually solved in conditions of limited technical and, especially, financial resources for its implementation. Therefore when developing monitoring system, it is necessary to strive for its optimality. It is necessary to find the optimal balance between the quantity of information, its representativeness and financial expenses.

Proposing a uniform definition reflecting the essence of the eutrophication process linked with monitoring and assessment system is not only an important starting point for developing a strategy for managing this process, but also for developing a methodology for assessing the trophic status of surface waters based on appropriate criteria [15].

Ecological diagnostics of aquatic systems requires multifactor assessments of trophic level and quality of water and the analysis of the most important relations between the components of ecosystems in order to find the dependency of trophic level on ecological, geographical factors and catchment management [16, 17]. Taking in account the above, it can be stated that indicators must meet the following demands:

- indicators reflect the state of biotic balance of water ecosystem,
- indicators are based on simple hydrochemical indices of water quality measured in any routine monitoring,

Table 6.3 Tasks of water monitoring

Problem	Detailing
What parameters, indicators, processes to observe (to measure)?	The choice of the list of measured parameters is characterized by a wide range of possible solutions and depends on the adopted methodology for assessing the trophic state
Where to locate observation points?	Determining the spatial discreteness of observations requires the information about the patterns of spatial distribution of water trophic state indicators and selection of points, that are the most informative for assessing spatial changes in this indicator
How often should observations be carried out?	The patterns of trophic state changes in indicators over time are theoretically well known. It is obvious that different parameters have different patterns of temporal changes, so the frequency of monitoring for them may be different. For example, the trophic state indicators during the vegetation season are characterized by high variability, therefore the required frequency of monitoring observations is not always possible to ensure
With what accuracy should the value of each parameter have to be measured?	The accuracy of the methods should be such, that the measurement error of eutrophication indicators should be many times smaller than the range of fluctuations of this indicator in the ecosystem,

- indicators can be used for express-monitoring of surface water and assessment of its ecological state,
- numerical indicators can be used for the purposes of mathematical modelling and solving engineering tasks,
- indicators can be the base for elaboration of ecological standards of water quality,
- indicators must characterized by low costs and simple calculations.

6.2 Limiting Factors—Nitrogen or Phosphorus?

To assess the production potential of waters the theoretical knowledge about the mechanisms of development of aquatic vegetation is very important. This potential is determined, in particular, by the availability of mineral nutrition elements, especially nitrogen and phosphorus. Resolution of the issue how and to what extent phytoplankton responds to nutrient loads and biogenic matter concentrations in water is fundamental for developing ideas about the mechanisms of surface water eutrophication. According to E. Odum, "the value of the concept of limiting factors, that are "functionally important" for an organism is what provides a starting point for ecological studies" [18].

In the second half of the XX century the so-called phosphorus paradigm (phosphorus limitation of phytoplankton in fresh waters) has emerged [19–21]. On the other hand, the research data from the last two or three decades have forced us to reconsider the established views. Under the influence of anthropogenic activities, natural biogeochemical elements cycles have undergone significant changes. According to estimates [22], starting since 1700, global changes in the amounts of C, N, P included in circulation lead to a significant increase in their runoff into water bodies due to changes in land use, fertilizer application and wastewater flows. During anthropogenic eutrophication and non-toxic water pollution, organic matter occurs the same processes as during the evolutionary development of lakes due to internal factors, but at higher rates [23]. The role of phosphorus as a limiting element no longer seems so indisputable and decisive. It turned out that nitrogen can also act as a limiting factor, both along with phosphorus (co-limiting) or even as a priority limiting factor. As follows from the results of numerous studies, the transition to limiting from phosphorus to nitrogen and vice versa depends on the geographical location of the reservoir, the characteristics of its catchment area, water body morphometry, and in close connection with biotic components of ecosystems [24].

However, the complete theoretical basis ideas about the nutrients limitation in water bodies appeared in connection with development of the fundamentals of E. Naumann regional limnology [25], who did not single out any one nutritional element as fundamental for the development of phytoplankton. His concept linked the productivity of phytoplankton with the geological conditions of the area, including the chemical composition of rocks and soils of the region. The hydrobiologist V. M. Rylov and the hydrochemist R. Mauha especially highly appreciated the leading principle of regional limnology—the dependence of the intensity of phytoplankton production development on the geological features of a particular area [26, 27].

Some authors pay attention to latitudinal variability of the limitation of water bodies by nutrients. Namely, in the European part of Russia, when moving from the north to the south there is a change in limitation from nitrogen to phosphorus. In other words, eutrophication in northern reservoirs is nitrogen limited, while in southern ones is phosphorus limited. So, in the tundra 81% of lakes are mostly nitrogen limited, in tajga there is about 25% of lakes is phosphorus limited, in forests—39% of the lakes is nitrogen limited. The number of lakes where productivity is limited by N and P, the greater, the lower their trophic level [28, 29].

However, the trophic status of the reservoir and the ratio of the concentrations of nitrogen, phosphorus and carbon are also important here. That's why it can be concluded that the nature of eutrophication limiting has the ambiguous character indicating frequent co-limitation with nitrogen and phosphorus.

According to Abel et al. [30] who studied the latitudinal variability of chlorophyll content in water bodies in connection with the concentration of nutrients, phosphorus is still more important as limiting factor, while nitrogen is statistically more significant only for shallow lakes. However, when developing practical recommendations for preventing eutrophication, it is important to take nitrogen into account.

Another factor influencing the eutrophication limitation is the specifics of the catchment area and specific types of land [31]. The authors state that the relationship between nitrogen, phosphorus and the level of phytoplankton productivity are largely regulated by the nature of the catchment area and the way of its development.

Other researches indicate that weather conditions can make an additional contribution: phytoplankton in agricultural landscapes is limited by phosphorus, but during drought the limitation shifts to nitrogen [32].

The depth of the lakes turned out to be one of the factors determining the nature of the limitation. It is believed that shallow-water lakes (with a depth of less than 6 m) are limited in nitrogen, and deep-water lakes are limited in phosphorus. Thus, total nitrogen turned out to be significant predictor of phytoplankton biomass in shallow lakes [33, 34].

Basing on the data from 2561 lakes Ward claims that the phosphorus content and climatic parameters (air temperature, precipitation, cloudiness, insolation) explain the chlorophyll content in 42% and 38%, respectively. So, it can be concluded that climatic features can act synergistically with phosphorus [35].

Basing on data for olitrophic Lake Sunapee (New Hampshire, USA) it was concluded that typical summer chlorophyll content (median) depends on climatic conditions, while the peak ("bloom") is determined by phosphorus load [35].

Using the example of three lakes in China of different trophic status, it was shown that the interaction of the temperature factor and the content of nutrients is determined by the taxonomic composition of the algal flora [36].

Due to the extraordinary complexity of the problem of establishing a priority limiting factor and the existence of a phosphorus-limiting paradigm up to that time, the question of the nature of the limitation of water eutrophication is not decided finally. The above mentioned facts allow us to conclude only that nitrogen, just like phosphorus, in many cases must be considered as a limiting element for phytoplankton. It is also necessary to abandon attempts to outline the boundary concentrations of nitrogen and phosphorus for water bodies in different regions as unpromising, as evidenced by all accumulated experience [24]. The same author suggests that there are probably no zonal factors of biogenic limitation of productivity, but in each water body there is a local set of conditions that ensure changing periods of nitrogen or phosphorus limitation of phytoplankton photosynthesis.

Anthropogenic eutrophication in fresh waters creates a highly unstable state with sharp fluctuations (violent "algal blooms" during which nutrients are released, causing another "bloom"). In this case, one of the substances cannot be put forward as a limiting factor. During such fluctuations, phosphorus, nitrogen, carbon dioxide and many other substances can quickly replace each other as limiting factors.

6.3 Assimilation Capacity—Comprehensive Approach Towards Considering Nutrients Loads

Undertakings related to preventing the negative effects of eutrophication are mainly conducted towards reducing the anthropogenic loads of nutrients discharged into surface waters from various sources. Numerous literature reports from different regions of the world indicate that the basic load of nitrogen to surface waters is supplied from dispersed sources (agriculture, atmospheric pollution), with the dominating share of area runoff from agricultural land, which provides about 64% of the total nitrates load. The main source of phosphorus in surface waters are municipal wastewater discharges which together with industrial wastewaters are responsible for the introduction of approx. 88% of total phosphate load into recipients [37–41].

The share of diffused and point sources of biogenic pollution in selected water bodies in Europe is illustrated in Table 6.4.

The share of different sources of nutrients supply in the Baltic Sea and Irish Sea are presented at Figs. 6.1 and 6.2.

Global nutrient transport in the XX century was dominated by agricultural sources, and from 1900 to 2000, nitrogen flow from this source increased from 19 to 51% of total nitrogen load and phosphorus flow from 35 to 56% of total phosphorus load. The highest increase in nitrogen and phosphorus global supply over this period was caused by point sources (mostly wastewater discharge), which increased from 2.0 to

Table 6.4 The share of point and area sources of biogenic pollution

Catchment	Nitrogen				Phosphorus			
	Total diffuse sources (%)	Therefrom agricultural sources (%)	Total point sources (%)	Therefrom urban wastewater (%)	Total diffuse sources (%)	Therefrom agricultural sources (%)	Total point sources (%)	Therefrom urban wastewater (%)
North Sea	69	85	31	89	44	85	56	77
Baltic Sea	45	70–90*	12	90	45	60–80[a]	20	90
Po River basin	80	–	20	57	93	–	7	90
Ebro river basin	64	–	–	–	–	–	88	–
Rhine River basin	69	–	31	84[b]	51	–	49	89
Danube River basin	84	42	16	25	67	28	33	51
Lake Constance	81	–	19	–	88	–	12	–

Note Not included in this table are natural background loadings and unspecified loads
[a] Agricultural contribution to total riverine load
[b] Diffuse urban loads included

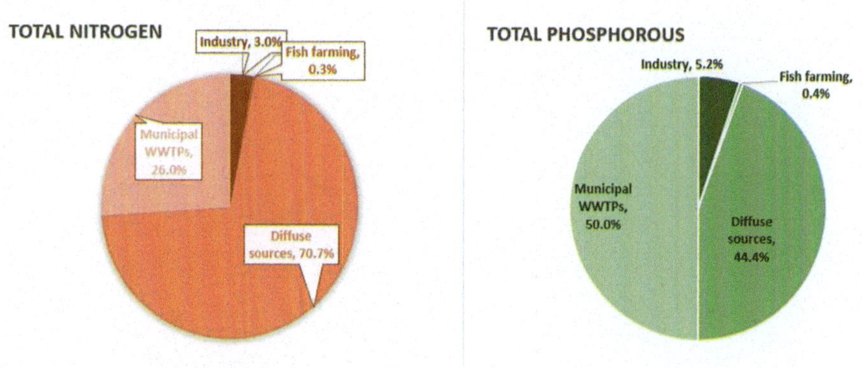

Fig. 6.1 The sources of biogenic loads in the Baltic Sea catchment [39]

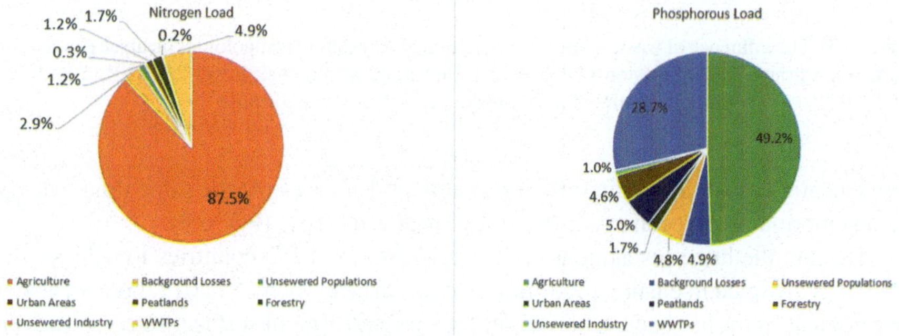

Fig. 6.2 The sources of biogenic loads in the Irish sea catchment [42]

8.0 TgN year^{-1} (340% increase) and from 0.2 to 1.0 TgP year^{-1} (500% increase), respectively [43]. The global forecast for future nutrient discharges indicates their increase from 6.4 TgN year^{-1} and 1.3 TgP year^{-1} in 2000 to 12.0–15.5 TgN year^{-1} and 2.4–3.1 TgP year^{-1} in 2050 [44].

The point sources, mainly municipal wastewaters, play an important role in eutrophication abatement strategies. The removal of nitrogen and phosphorus compounds from wastewaters became a key component in eutrophication combating strategies, based on legal requirements for treated wastewater discharge into aquatic environment introduced in all European and many non-European countries. These requirements are systematically becoming stricter in terms of nutrient content which has resulted in the use of advanced nutrient removal technologies in complex and expensive wastewater treatment systems. Thus, the problem of establishing a reasonable level of nutrient removal from wastewater, justified from an ecological and economic point of view is a very important issue. When the quality of treated wastewater approaches the quality of unpolluted recipients, the treatment costs increase

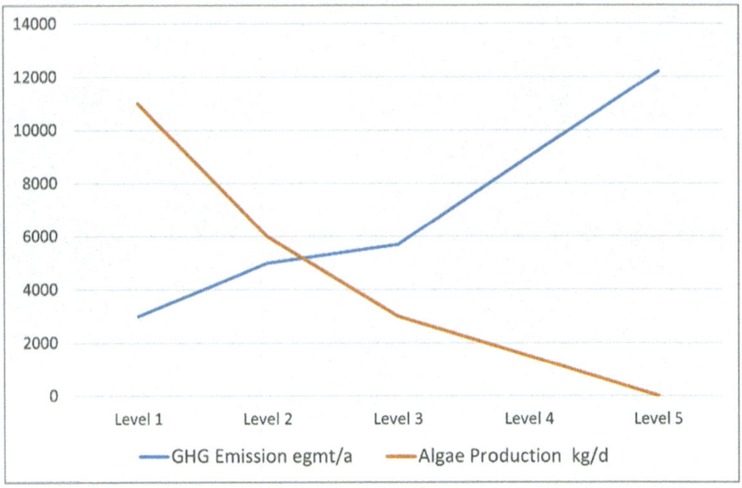

Fig. 6.3 The influence of wastewater discharge treated in different technologies on algae production and appropriate GHG emission [45]. Level 1—no nutrient removal; Level 2—BNR; Level 3—EBNR; Level 4—limit of treatment technology; Level 5—reverse osmose

exponentially and the profitability of treatment decreases. At the same time, energy consumption and greenhouse gas emissions also increase (Fig. 6.3).

Despite the huge investment outlays and efforts of EU countries to achieve the WFD goal aimed to ensure good status for all surface waters by 2015, good or better ecological status has been achieved for only around 40% of surface waters, however, the weak ecological status was determined mainly by high trophic state of waters (Fig. 6.4).

Surface water bodies in north-western Europe have the lowest status. In Belgium (Flanders), northern Germany and the Netherlands the ecological status of more than 90% of surface waters is reported to be less than good. Other problem areas include Czechia, southern England (United Kingdom), northern France, southern Germany, Hungary and Poland, as well as several individual RBDs (River Basin Districts) in some countries where 70–90% of surface water bodies are reported to have less than good status [46].

Focusing only on limiting the nutrients supply from point sources (mainly wastewaters) and ignoring significant area sources, leads to unjustified investments in wastewater treatment and ineffectiveness of eutrophication preventing strategies. This is confirmed by the unsatisfactory results of the practice of implementing of the Water Framework Directive requirements in terms of achieving a good water status in member states. According to European Commission [47] only 40% of surface waters in EU member countries achieved good ecological status and 33%-achieved good chemical status, meanwhile, the capital investment costs of achieving good water status are significant and amounted to at least 142 billion EURO Fig. 6.5.

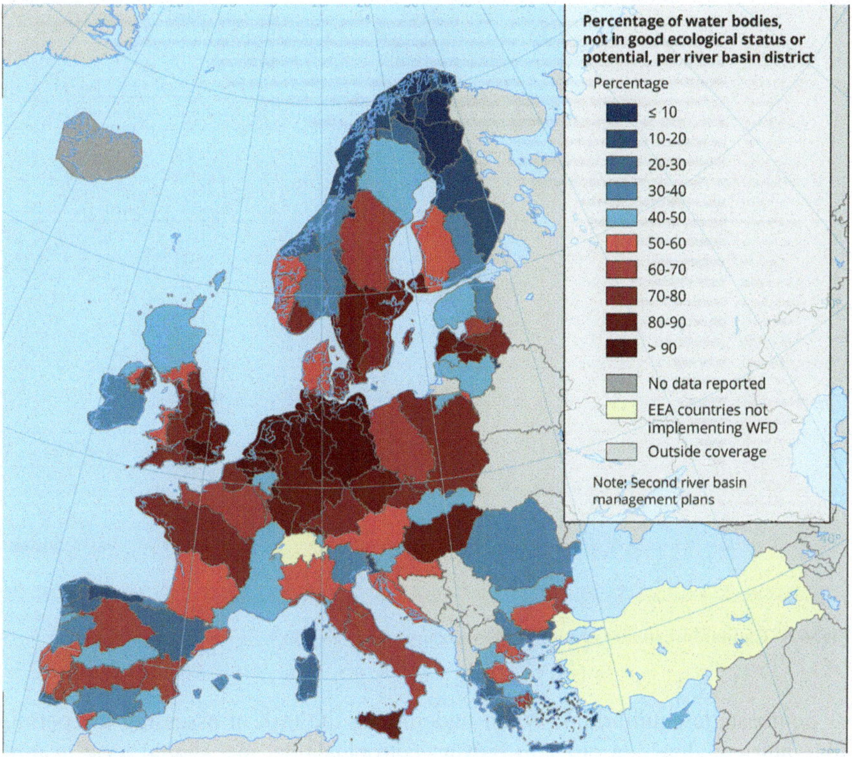

Fig. 6.4 Surface water status in Europe ([46] CC BY)

The structure of costs allocated to achieving WFD goals indicates that in many countries the dominant part of funds is allocated to the elimination of pollution discharged from point sources, e.g.: in Belgium—90%, in Portugal—75.9%, in Romania—79%, in Cyprus—48%, in Slovakia—40%, in Hungary—29% [47].

Regardless of the broad-scale projects implemented in basin countries, aimed at reduction of biogenic loads discharged to surface waters mainly from point sources (excluding practically the diffused sources) and related huge capital expenditures, the expected effects in eutrophication abatement were not achieved and cost–benefit ratio is rather high and unprofitable. It seems that there are two basic reasons for failure: (1) the lack of an appropriate system of regional ecological standards and establishment of permissible loads from all biogenic sources according to assimilative capacity of water body, (2) the lack of trophic status assessment approaches based on integrated indicators reflecting the biotic balance of water ecosystem, (3) the lack of an effective monitoring and control strategy of nutrient loads from all sources.

The term "assimilative capacity" used in water sciences has never been strictly defined and different authors give varying definitions [48]. In general, the assimilative capacity of water body defined there as its ability to accept anthropogenic pollutants loads from all sources without the disturbing the ecosystem functioning within

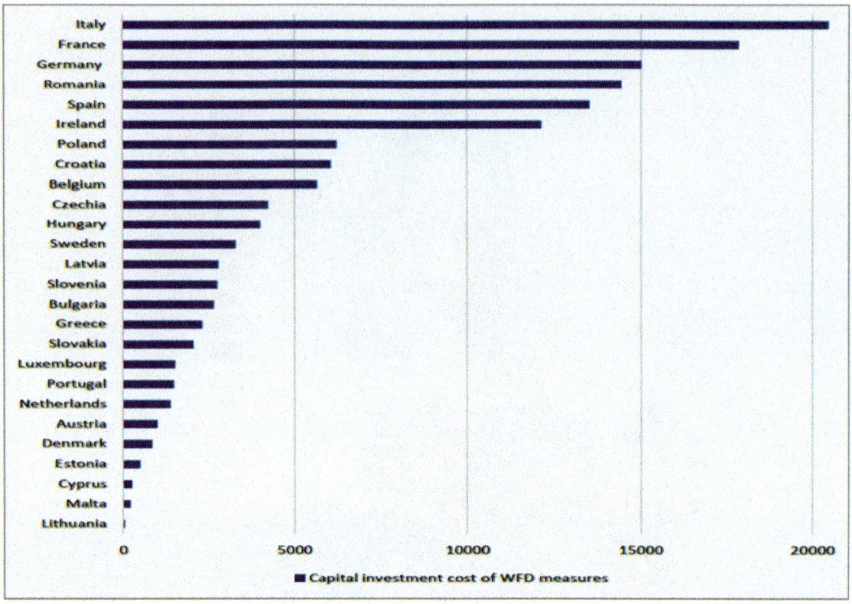

Fig. 6.5 Capital investment cost of WFD measures ([47] CC BY)

its homeostatic equilibrium. When undertaking the task of assessing the permissible nutrients loads in order to prevent eutrophication, one should have in mind the ecosystem's ability to bear nutrient loads without disturbing of biotic balance, manifested in the trophic level increase. The calculation of assimilation capacity is very complex task and depends on biotic and abiotic features of ecosystem, climate conditions and qualitative and quantitative characteristics of anthropogenic impacts.

The development of permissible loads of biogenic substances should include the calculation of the amounts of allowable loads for each source, and their sum cannot exceed the established assimilative capacity of the recipient. Based on the assimilative capacity of the recipient and the number of potential sources of pollution, the amounts of discharged loads should be accordingly determined for each of them. The concept of assimilative capacity of nutrients loads management considering all point and area sources of nutrients supply to water recipient is presented at Fig. 6.6.

Generally speaking, anthropogenic impacts on any natural water ecosystems should not exceed their ability to compensate for these impacts in self-purification processes within their homeostasis balance. So, the civilization development must be ensured within ecological capacity of natural ecosystems.

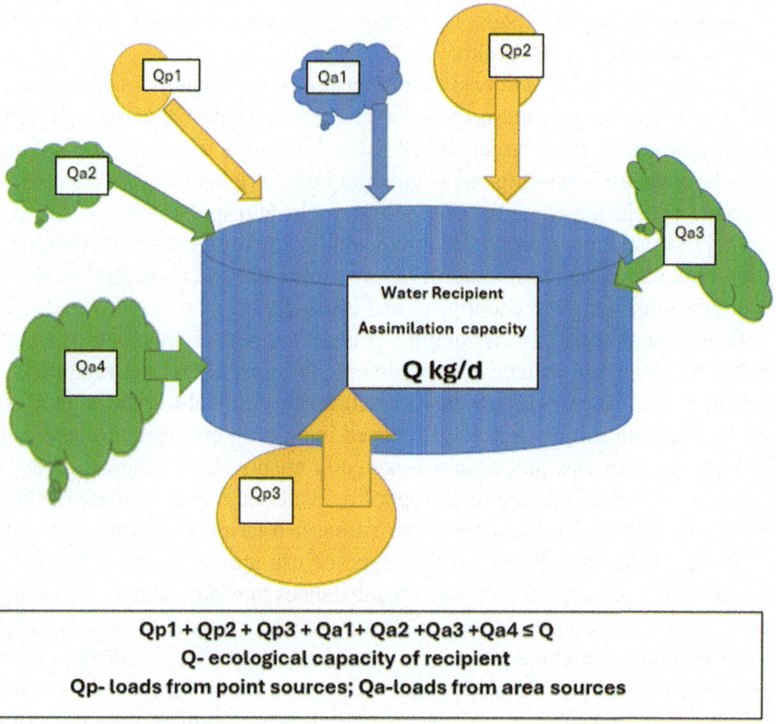

Fig. 6.6 Concept of assimilative capacity approach for evaluation of permissible loads

6.4 Concept of Ecological Rating

Environmental quality is characterized by physical, chemical, biological indicators or by combination of them. A favourable environment is understood as an environment which quality ensures the sustainable functioning of natural ecosystems. Environmental standards are established for environment characteristics that ensure the sustainable functioning of natural ecosystems and biological diversity. Standardization of environmental quality is an activity that is carried out in 2 directions: (1) development of environmental quality standards (physical, chemical, biological and other indicators, which compliance ensures a favourable environment quality) and (2) development of standards for permissible impact on the environment (indicators of the impact of anthropogenic activities that comply with environmental quality standards). The standards for permissible anthropogenic load have to be established in accordance with the magnitude of the permissible cumulative impact of all sources/activities on the environment within specific territories and water areas and ensure the sustainable functioning of natural ecological systems.

Standards underlie the balance of human environmental and economic interests. A measure of a reasonable combination of interests is the maximum permissible levels of anthropogenic loads, the excessing of which creates a danger to the natural environment and human health.

These standards are the basis for determining the legality of the economic entities activity, the effectiveness of observance of environmental and legal regulations, the economics of environmental activities. Indicators of environmental quality are also the basis for observing environmental human rights, conducting environmental assessments, the degree of environmental and legal responsibility, assessing environmental risks and much more. The main and final goal of environmental regulation is a scientifically based combination of environmental and economic interests, a kind of compromise between economics and ecology.

When standardizing water quality, it is necessary to remember, that water is not only a resource, but aquatic ecosystems are the most important regulators of living conditions on the planet. However, the currently applicable water quality standards are typically anthropocentric and focused mainly on ensuring the health safety for humans and fisheries and do not practically ensure the ecological safety of water systems. Consumer requirements regarding water quality depend on the purposes of water use: drinking water supply, recreation, fish farming irrigation and others.

Water quality standards for each type of use are set in the form of maximum permissible concentrations of various substances in water, which is to ensure its safe use for specific purposes. However, the nature of the impact of various substances in water on humans and the entire ecosystem may vary. These standards are the basis for decisions making, developing activities and estimating costs for environmental protection and fees for its use. In this regard, it is extremely important to establish scientifically based, carefully verified and achievable standards.

Tightening requirements for water quality provoke their non-compliance and lead to the accrual of huge fines for excess discharges of pollution. Therefore, it is necessary to develop such standards that will provide a balanced solution to environmental protection and economic development.

A compromise between nature and human can only be found, when anthropogenic loads do not exceed the compensatory capabilities of ecosystems. From this it can be concluded, that human well-being must be ensured within the capacity of biosphere, which exceeding by anthropogenic loads will lead to global degradation of biospheric conditions. Therefore, environmental protection should not be limited only to finding technologies to minimize them. Balancing the economic activities with nature's ability to cope with anthropogenic pressures—it is the main task of environmental protection. Appropriate environmental quality standards should serve as a tool.

Unfortunately, the environmental quality standards in force in many countries do not ensure effective protection of aquatic ecosystems, and a paradoxical situation arises: standards are becoming more restrictive, fees and costs are increasing, but the state of water bodies is not improving. One of the illustration of this situation is the ineffective actions under the Water Framework Directive.

There are several points of view on approaches and methodology for the development of environmental standards: based on a sanitary and hygienic approach to protect human health or based on the technological capabilities of BAT (Best Available Technologies). As practice shows, none of these approaches guarantees effective water protection, especially in the case of protection against eutrophication. There is a need for standards developed on the base of ecosystem approach, ensuring the ecological safety of the entire ecosystem, which in turn can also ensure good water quality, and in many cases are more profitable from an economic point of view. Taking into account the fact that the functioning of separate ecosystems under the same anthropogenic loads significantly depends on natural factors totality and local/regional conditions, ecological standards should be local/regional. Defining the ecological standards for the aquatic environment, it can be stated, that it is permissible concentrations of pollutants that do not disturb the homeostatic self-regulation mechanisms of water ecosystems. EPC (ecologically permissible concentrations) of substances may serve as the basis for determining the ecologically permissible loads (EPL) for aquatic ecosystems, which will not exceed their assimilation capacity.

Ecological standardization should be a part of environmental protection programs ensuring the ecological safety of aquatic ecosystems, but this requires the development and implementation of new approaches and methods for assessing ecological capacity, which in turn requires further research and overcoming administrative, legal and financial barriers.

6.5 Eutrophication Potential of Wastewaters—An Overlooked Aspect

Municipal wastewaters characterized by a great loads of nitrogen and phosphorus compounds bringing to the intensification of eutrophication process. The actions to reduce these loads focus on the reduction of nutrients content in treated wastewaters in accordance with strict legal requirements achievable only in advanced technologies. The technologies of enhanced biological nutrient removal (EBNR) can meet the appropriate legal requirements only if based on complicated energy-consuming and expensive treatment technological schemes and do not always guarantee effective protection against eutrophication and its consequences [49].

The legal requirements concerning nitrogen and phosphorus content in treated wastewater are actually developed mainly for total nitrogen and total phosphorus [50]. But the degree of influence of treated wastewaters discharged into the recipient on eutrophication development is determined not so much by the content of total nitrogen and total phosphorus, but by their bioavailable mineral forms directly available to aquatic vegetation, which was defined by the authors as "wastewater eutrophication potential" [49].

The classification of the content of bioavailable forms of phosphorus and nitrogen in treated wastewater and their "reactivity" is shown in Figs. 6.7 and 6.8.

The share of bioavailable phosphorus and nitrogen compounds in treated wastewaters varies and depends mostly on raw wastewater quality and treatment technologies

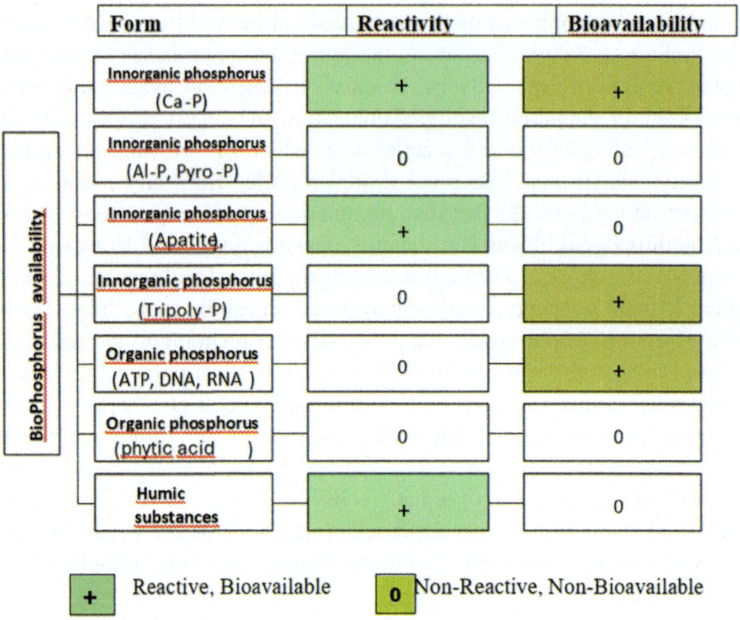

Fig. 6.7 Different forms of phosphorus in wastewater [51]

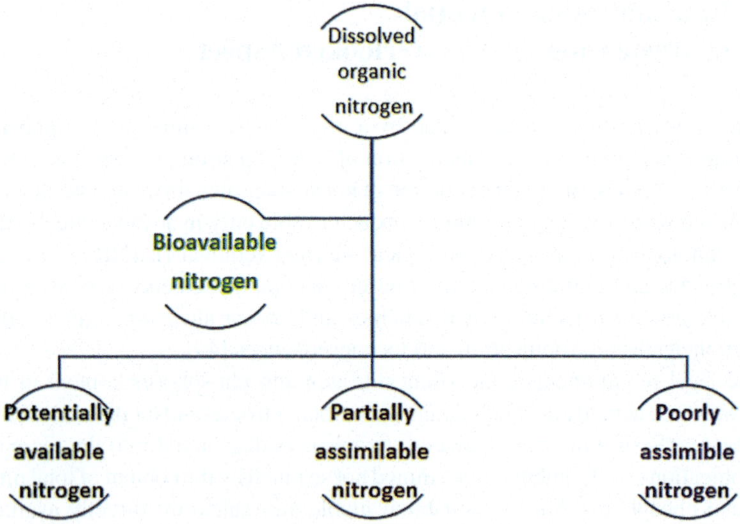

Fig. 6.8 Different forms of nitrogen in wastewater [52]

used. In most cases, more than 50% of phosphorus in treated sewage is unavailable to aquatic vegetation [53]. Research results even prove, that raw sewage contains fewer forms of phosphorus available to aquatic vegetation than treated sewage, and their share is determined by the technology of biogenic substances removal [54].

In the total load of nitrogen compounds introduced into surface waters, the share of bioavailable forms is approximately 20–40% [17]. Other sources report, that the share of bioavailable nitrogen in the total load introduced into rivers is 12–45%. Depending on the treatment technology, the share of bioavailable organic nitrogen in wastewater may range from 4 to 40% [55]. According to other literature reports, the share of potentially bioavailable organic nitrogen in treated sewage was 2–16% [56]. The bioavailability of organic nitrogen introduced into recipients will also largely depend on the conditions of biochemical decomposition of organic substances in water.

However, the analysis of knowledge about the mechanisms of the eutrophication process, its main factors and the role of nitrogen and phosphorus in the development of aquatic vegetation allowed the conclusion, that among all forms of nitrogen and phosphorus compounds in water (dissolved inorganic and organic as well as organic and inorganic suspensions), only dissolved inorganic nutrients forms are directly available to aquatic vegetation. Therefore it is their content that determines the eutrophication potential of treated wastewaters. Consideration of this knowledge on the share of bioavailable forms of nitrogen and phosphorus in wastewaters constitutes the basis for the selection of appropriate wastewater treatment technologies, ensuring a reduction of their impact on the development of eutrophication. The advanced methods currently used to eliminate nutrients from wastewater in accordance with applicable requirements are based on complex and expensive technological systems. Unfortunately, they do not always provide effective protection against eutrophication and its effects, because they usually focus on the removal of general forms of nutrients, without taking into account the eutrophication potential of treated wastewater, determined by the content of their bioavailable forms.

The authors of this monograph conducted research on the eutrophication potential of wastewaters discharged from selected treatment plants, assessing this potential on the base of the content of bioavailable (mineral forms) of biogenic compounds directly accessible to aquatic vegetation. The results of analysis (Table 6.5) confirms the fact, that in terms of bioavailability of nutrients biologically treated wastewater has a greater eutrophication potential than before treatment, which was also confirmed by other researchers [17, 49, 57, 58].

The currently applicable quality standards for treated wastewater introduced into the aquatic environment are based mainly on the content of total phosphorus and total nitrogen, without taking into account the content of their bioavailable forms after biological treatment. And only some countries set permissible values for some mineral forms of nutrients. This approach does not provide effective protection of wastewater recipients against eutrophication and requires the use of extremely expensive technologies. However, taking into account the eutrophication potential of wastewaters it is possible to select ecologically and economically justified technologies for the elimination of nutrients [50].

Table 6.5 Content of bioavailable forms of phosphorus and nitrogen in raw and treated sewage

WWTP	Bioavailable N (%)	Growth of bioavailable N after purification (%)	Bioavailable P (%)	Growth of bioavailable P after purification (%)
Krakow-Płaszow (PL)				
Raw wastewater	70.01	9.86	51.40	3.72
Treated wastewater	79.87		55.12	
Helsinki (FIN)				
Raw wastewater	73.80	−7.70	46.59	−1.14
Treated wastewater	66.10		45.45	
Saint Petersburg Central Aeration Station (RUS)				
Raw wastewater	49.12	22.72	36.96	20.18
Treated wastewater	71.84		57.14	
Saint Petersburg South-West WWTP (RUS)				
Raw wastewater	56.85	19.27	51.40	5.12
Treated wastewater	76.12		56.52	
Chicago-Hanover Park (USA)				
Raw wastewater	65.95	25.35	48.89	47.41
Treated wastewater	91.30		96.30	
Chicago-Calumet (USA)				
Raw wastewater	63.58	23.14	62.00	32.44
Treated wastewater	86.72		94.44	
Chicago-Kirie (USA)				
Raw wastewater	63.03	30.53	45.00	42.50
Treated wastewater	93.56		87.50	
Chicago-Egan (USA)				
Raw wastewater	61.64	33.62	49.12	43.98

(continued)

Table 6.5 (continued)

WWTP	Bioavailable N (%)	Growth of bioavailable N after purification (%)	Bioavailable P (%)	Growth of bioavailable P after purification (%)
Treated wastewater	95.26		93.10	
Chicago-North Side (USA)				
Raw wastewater	68.69	21.00	45.16	47.70
Treated wastewater	89.69		92.86	
City of Penticton (CAN)				
Raw wastewater	48.13	41.03	36.73	−3.40
Treated wastewater	89.16		33.33	

Extensive research on the eutrophication potential of wastewaters treated in most common technologies allowed proposing the original approach to appropriate selection of wastewater technology. It is based on application of computer simulation in order to define the proper wastewater treatment technology, which ensure the minimum eutrophication potential of treated wastewater and will be economically justified. The aim of computer simulation was to determine the efficiency of removal of bioavailable forms of biogenic compounds in 10 most common wastewater treatment technologies: traditional secondary treatment, Anoxic–Oxic technology (AO), Bardenpho technology (3-stage), modified Bardenpho technology (5-stage), University of Cape Town technology (UCT), modified University of Cape Town technology (MUCT), Anaerobic-Oxic technology (A/O),Oxic-Anoxic technology (OA), Johannesburg technology (JHB) and MJHB (modified JHB) technology. The results of simulation research are presented in Table 6.6, presenting the content of total nitrogen, total phosphorus, dissolved inorganic nitrogen, dissolved inorganic phosphorus) and the percentage share of mineral nitrogen and phosphorus in their total content in treated wastewaters on the basae of data obtained for 10 WWTP [50].

Analysis of the results of simulation of the content of various forms of nitrogen and phosphorus in wastewater treated in different technologies shows, that none of the 10 systems provides required quality standards for treated wastewater. The share of bioavailable forms of nitrogen and phosphorus in total content of this nutrients oscillates between 90–94% and 85–95% respectively. The most effective elimination of nutrients was reached in JHB and MJHB technologies. Simultaneously in the same technological systems, the lowest inorganic phosphor and inorganic nitrogen content was also reached. Wherein, the lowest eutrophication potential has been achieved by wastewater treated in JHB and MJHB technologies.

Simulation research on eutrophication potential of treated wastewaters discharged into water recipients allowed the authors to develop the decision-making algorithm

Table 6.6 Bioavailable nutrients content in treated wastewater

Technology	TN (mg/l)	TP (mg/l)	DIN (mg/l)	DIP (mgl)	DIN (%)	DIP (%)
Traditional	58.15	2.76	54.66	2.61	94	95
AO	54.08	2.77	50.89	2.62	94	95
Bardenpho-3	53.26	2.24	49.85	2.07	94	92
Bardenpho-5	53.89	2.48	50.73	2.32	94	94
JHB	33.1	1.39	29.71	1.18	90	85
MJHB	33.09	1.39	29.71	1.19	90	85
UCT	39.16	1.93	35.83	1.74	91	90
MUCT	41.05	1.85	37.81	1.66	92	90
A/O	35.32	1.79	31.85	1.60	90	90
OA	56.59	2.83	53.18	2.68	94	95

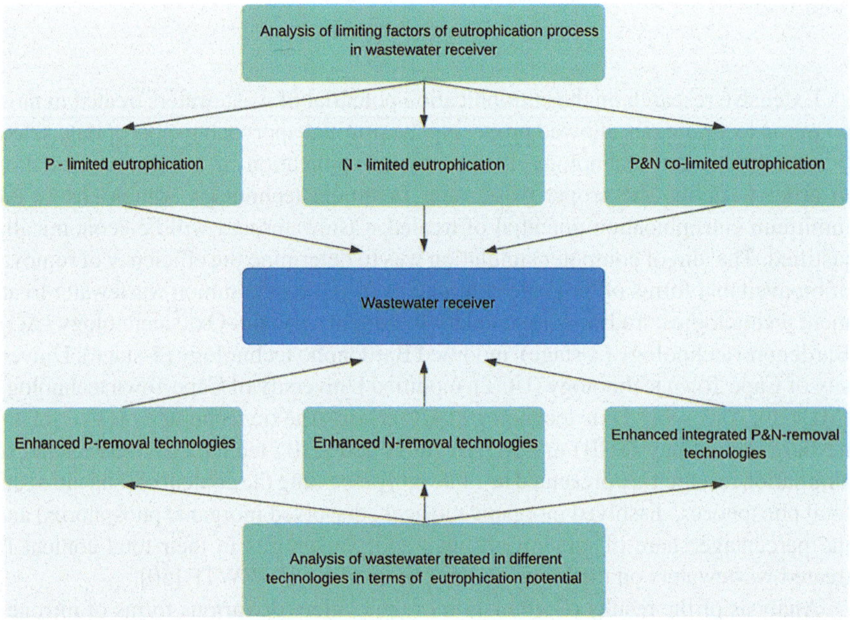

Fig. 6.9 The recipient-oriented approach for selecting the optimal wastewater treatment technology [59]

for selecting the optimal wastewater treatment technology, taking into account their impact on the intensification of eutrophication processes in recipients. In order to improve surface water state by ensuring an optimal quality of treated wastewater, a dual recipient-oriented approach has been developed. The conceptual scheme of the proposed approach is shown in Fig. 6.9.

References

1. Neverova-Dziopak E (2009) Problems of ecological monitoring of surface waters. Environ Prot Eng 35(4):81–91
2. Neverova-Dziopak E (2021) Surface water eutrophication in Poland: assessment and prevention. In: Quality of water resources in Poland. Springer, pp 321–346. https://doi.org/10.1007/978-3-030-64892-3_14
3. EC (1979) Council Directive 79/923/EEC of 30 October 1979 on the quality required of shellfish waters
4. EC (1991) Council Directive 91/676/EEC of 12 December 1991 concerning the protection of waters against pollution caused by nitrates from agricultural sources
5. EC (1991) Council Directive 91/271/EEC of 21 May 1991 concerning urban waste-water treatment
6. EC (1992) Council Directive 92/43/EEC of 21 May 1992 on the conservation of natural habitats and of wild fauna and flora
7. EC (2000) Directive 2000/60/EC of the European Parliament and of the Council of 23 October 2000 establishing a framework for Community action in the field of water policy
8. EC (2006) Directive 2006/44/EC of the European Parliament and of the Council of 6 September 2006 on the quality of fresh waters needing protection or improvement in order to support fish life
9. EC (2011) Directive 2006/11/EC of the European Parliament and of the Council of 15 February 2006 on pollution caused by certain dangerous substances discharged into the aquatic environment of the community
10. Soszka H (2009) Methodical problems associated with the assessment of lake eutrophication for designation of zones vulnerable to nitrates water framework directive. Woda-Środowisko-Obszary Wiejskie 9(1):151–159
11. EC (2009) Guidance document on eutrophication assessment in the context of European water policies. Guidance document No 23. Publications Office. https://doi.org/10.2779/49795
12. Savanina Y, Barsky E, Fomina I, Lobakova E (2019) Monitoring eutrophication pollutions using electrochemical indicators. Inf Technol Sci Educ Manag 4
13. Edelshtein KK, Datsenko YuS, Puklakov VV (2021) Water blooming in the reservoir with long-term discharge regulation. Water Resour 48(2):245–253. https://doi.org/10.1134/S0097807821020044
14. Edelshtein KK, Grechushnikova MG, Datsenko YuS, Puklakov VV (2012) Diagnostic modelling of within-water processes in reservoirs. Water Resour 39(4):432–445. https://doi.org/10.1134/S0097807812040045
15. Andersen JH, Schlüter L, Ærtebjerg G (2006) Coastal eutrophication: recent developments in definitions and implications for monitoring strategies. J Plankton Res 28(7):621–628. https://doi.org/10.1093/plankt/fbl001
16. EC (2000) Towards environmental pressure indicators for the EU
17. Neverova-Dziopak E (2010) Podstawy zarządzania procesem eutrofizacji antropogenicznej. Wydawnictwa AGH
18. Odum E (1971) Fundamentals of ecology, 3rd edn. W.B. Saunders Co., Philadelphia
19. Lee GF, Jones RA (1984) Summary of US OECD eutrophication study. Results and their application to water quality management. SIL Proc 1922–2010 22(1):261–267. https://doi.org/10.1080/03680770.1983.11897299
20. Redfield AC (1934) On the proportions of organic derivatives in sea water and their relation to the composition of Plankton. James Johnstone Memorial Volume
21. Redfield AC (1960) The biological control of chemical factors in the environment. Sci Prog 11:150–170
22. Mackenzie F, Ver L, Lerman A (2002) Century-scale nitrogen and phosphorus controls of the carbon cycle. Chem Geol 190:13–32. https://doi.org/10.1016/S0009-2541(02)00108-0
23. Alimov AF (2003) Towards a theory of the functioning of aquatic ecosystems. Backhuys

24. Rizhinashvili AL (2022) An outline of the theory of the functioning of aquatic ecosystems: nutrient limitation. Biol Bull Rev 12(6):596–608. https://doi.org/10.1134/S2079086422060068
25. Naumann E (1933) Grundzüge der regionalen Limnologie. Nature 131(3303):223–224. https://doi.org/10.1038/131223a0
26. Maucha R (1924) Upon the influence of temperature and intensity of light on the photosynthetic production of nannoplankton. SIL Proc 1922–2010 2(1):381–401. https://doi.org/10.1080/03680770.1924.11898327
27. Rylov W (1929) Bemerkungen betreffs des regionallimnologischen Studiums. Verhandlungen Der Int Vereinigung Für Theoretische Und Angewandte Limnol 4
28. Gashkina NA (2011) Zonal features of the distribution of biogenic elements and organic matter in small lakes. Water Resour 38(3):352–371. https://doi.org/10.1134/S0097807811010039
29. Gashkina NA, Moiseenko TI (2010) Trophicity limitation in small lakes by main nutrients. Dokl Earth Sci 435(1):1539–1543. https://doi.org/10.1134/S1028334X10110280
30. Abell JM, Özkundakci D, Hamilton DP, Jones JR (2012) Latitudinal variation in nutrient stoichiometry and chlorophyll-nutrient relationships in lakes: a global study. Fundam Appl Limnol 181(1):1–14. https://doi.org/10.1127/1863-9135/2012/0272
31. Vanni MJ, Renwick WH, Bowling AM, Horgan MJ, Christian AD (2011) Nutrient stoichiometry of linked catchment-lake systems along a gradient of land use. Freshw Biol 56(5):791–811. https://doi.org/10.1111/j.1365-2427.2010.02436.x
32. Hayes NM, Vanni MJ, Horgan MJ, Renwick WH (2015) Climate and land use interactively affect lake phytoplankton nutrient limitation status. Ecology 96(2):392–402. https://doi.org/10.1890/13-1840.1
33. Donis D, Mantzouki E, McGinnis DF, Vachon D, Gallego I, Grossart H, de Senerpont Domis LN, Teurlincx S, Seelen L, Lürling M, Verstijnen Y, Maliaka V, Fonvielle J, Visser PM, Verspagen J, van Herk M, Antoniou MG, Tsiarta N, McCarthy V, Ibelings BW (2021) Stratification strength and light climate explain variation in chlorophyll a at the continental scale in a European multilake survey in a heatwave summer. Limnol Oceanogr 66(12):4314–4333. https://doi.org/10.1002/lno.11963
34. Shuvo A, O'Reilly CM, Blagrave K, Ewins C, Filazzola A, Gray D, Mahdiyan O, Moslenko L, Quinlan R, Sharma S (2021) Total phosphorus and climate are equally important predictors of water quality in lakes. Aquat Sci 83(1):16. https://doi.org/10.1007/s00027-021-00776-w
35. Ward NK, Steele BG, Weathers KC, Cottingham KL, Ewing HA, Hanson PC, Carey CC (2020) Differential responses of maximum versus median chlorophyll-*a* to air temperature and nutrient loads in an oligotrophic lake over 31 years. Water Resour Res 56(7). https://doi.org/10.1029/2020WR027296
36. Dong J, Zhou Q, Gao Y, Gu Q, Li G, Song L (2018) Long-term effects of temperature and nutrient concentrations on the phytoplankton biomass in three lakes with differing trophic statuses on the Yungui Plateau, China. Ann Limnol Int J Limnol 54:9. https://doi.org/10.1051/limn/2017031
37. Baer J, Eckmann R, Rösch R, Arlinghaus R, Brinker A (2017) Managing upper lake Constance fishery in a multi-sector policy landscape: beneficiary and victim of a century of anthropogenic trophic change, pp 32–47
38. Hardenbicker P, Weitere M, Ritz S, Schöll F, Fischer H (2016) Longitudinal plankton dynamics in the rivers Rhine and Elbe. River Res Appl 32(6):1264–1278. https://doi.org/10.1002/rra.2977
39. HELCOM (2009) Eutrophication in the Baltic Sea
40. Pirrone N, Trombino G, Cinnirella S, Algieri A, Bendoricchio G, Palmeri L (2005) The driver-pressure-state-impact-response (DPSIR) approach for integrated catchment-coastal zone management: preliminary application to the Po catchment-Adriatic Sea coastal zone system. Reg Environ Change 5(2–3):111–137. https://doi.org/10.1007/s10113-004-0092-9
41. Torrecilla NJ, Galve JP, Zaera LG, Retamar JF, Álvarez ANA (2005) Nutrient sources and dynamics in a Mediterranean fluvial regime (Ebro river, NE Spain) and their implications for water management. J Hydrol 304(1–4):166–182. https://doi.org/10.1016/j.jhydrol.2004.07.029
42. O'Boyle S, Wilkes R, Mcdermott G, Longphuirt S (2015) Transitional and coastal waters: in water quality in Ireland 2010–2012, pp 110–139

43. Beusen AHW, Bouwman AF, Van Beek LPH, Mogollón JM, Middelburg JJ (2016) Global riverine N and P transport to ocean increased during the 20th century despite increased retention along the aquatic continuum. Biogeosciences 13(8):2441–2451. https://doi.org/10.5194/bg-13-2441-2016
44. Van Drecht G, Bouwman AF, Harrison J, Knoop JM (2009) Global nitrogen and phosphate in urban wastewater for the period 1970 to 2050. Glob Biogeochem Cycl 23(4). https://doi.org/10.1029/2009GB003458
45. Falk MW, Neethling J, Reardon DJ (2011) Striking the balance between nutrient removal in wastewater treatment and sustainability. Water Intell Online 10:9781780403298. https://doi.org/10.2166/9781780403298
46. EEA (2018) European waters—assessment of status and pressures 2018, EEA report, 7/2018
47. EC (2021) Economic data related to the implementation of the WFD and the FD and the financing of measures—final report. Publications Office. https://doi.org/10.2779/163850
48. Campbell IC (1981) A critique of assimilative capacity. Water Pollut Control Fed 53(5)
49. Preisner M, Neverova-Dziopak E, Kowalewski Z (2020) Analysis of eutrophication potential of municipal wastewater. Water Sci Technol 1–10. https://doi.org/10.2166/wst.2020.254
50. Kowalewski Z, Neverova-Dziopak E, Preisner M (2016) Zastosowanie symulacji komputerowej procesu osadu czynnego do minimalizowania potencjału eutrofi zujacego scieków oczyszczonych. Ochrona Srodowiska 38(3):23–28
51. Venkiteshwaran K, McNamara PJ, Mayer BK (2018) Meta-analysis of non-reactive phosphorus in water, wastewater, and sludge, and strategies to convert it for enhanced phosphorus removal and recovery. Sci Total Environ 644:661–674. https://doi.org/10.1016/j.scitotenv.2018.06.369
52. Fan L, Brett MT, Li B, Song M (2018) The bioavailability of different dissolved organic nitrogen compounds for the freshwater algae Raphidocelis subcapitata. Sci Total Environ 618:479–486. https://doi.org/10.1016/j.scitotenv.2017.11.096
53. Li B, Brett MT (2012) The impact of alum based advanced nutrient removal processes on phosphorus bioavailability. Water Res 46(3):837–844. https://doi.org/10.1016/j.watres.2011.11.055
54. Ekholm P, Krogerus K (2003) Determining algal-available phosphorus of differing origin: routine phosphorus analyses versus algal assays. Hydrobiologia 492(1–3):29–42. https://doi.org/10.1023/A:1024857626784
55. Czerwionka K (2016) Influence of dissolved organic nitrogen on surface waters. Oceanologia 58(1):39–45. https://doi.org/10.1016/j.oceano.2015.08.002
56. Stepanauskas R (2000) Bioavailability and sources of DOC and DON in macrophyte stands of a tropical coastal lake. Hydrobiologia 436(1/3):241–248. https://doi.org/10.1023/A:1026537905947
57. Graham M, Blignaut J, Villiers L, Mostert D, Sibande X, Gebremedhin S, Harding B, Rossouw N, Freese N, Ferrer S, Browne M (2012) Development of a generic model to assess the costs associated with eutrophication
58. Michalski M, Conroy N (1972) The "oligotrophication" of Little Otter Lake, Parry Sound District—1971–1972. Ontario Ministry of the Environment
59. Preisner M, Neverova-Dziopak E, Kowalewski Z (2020) Mitigation of eutrophication caused by wastewater discharge: a simulation-based approach. Ambio. https://doi.org/10.1007/s13280-020-01346-4

Chapter 7
Original Concept of Eutrophication Management

Abstract This chapter presents the author's original, multi-aspect concept of managing the eutrophication process, taking into account those aspects and problems, that were identified and analysed in the monograph and which must be taken into account when developing a strategy for managing this process at various levels. Starting from the basic assumptions of the functional direction of the system approach, the authors developed the holistic multilevel concept of eutrophication process managing, which takes into account the aspects currently ignored in routine practices of this process managing. The concept covers coordinated activities' totality at various hierarchical levels of process management.

Keywords Multilevel concept · Theoretical knowledge · Process mechanisms · Legal regulations · Organizational, technical and technological solutions

7.1 Milestones of Eutrophication Management Strategies

Eutrophication is still a problem worldwide. It would seem that all the causes and consequences of eutrophication are known. The European Environment Agency's (EEA) assessment shows, that it still remains a large scale unsolved problem (Figs. 7.1 and 7.2). During the last decades, the implemented legal regulations in EU member countries (WFD) did not lead to satisfying effect on the improvement of surface water state [1].

The question is: why, despite many years of research on eutrophication, constant enrichment of knowledge on this subject and huge financial outlays on water protection and reclamation, this problem has not been solved so far? There is no clear and monosyllabic answer to this question. The problem is multi-level and multi-factorial, and there is no universal solution. There are many existing barriers to solving it: scientific, methodological, technological, organizational, legislative, political, economic, educational, environmental awareness and education and others.

It is necessary to apply modern knowledge about the impact of human activities on aquatic ecosystems to solve the problem of eutrophication. Until now, the main focus

E. Neverova-Dziopak and Z. Kowalewski, *Eutrophication: A Global Environmental Problem*, Springer Water, https://doi.org/10.1007/978-3-031-83926-9_7

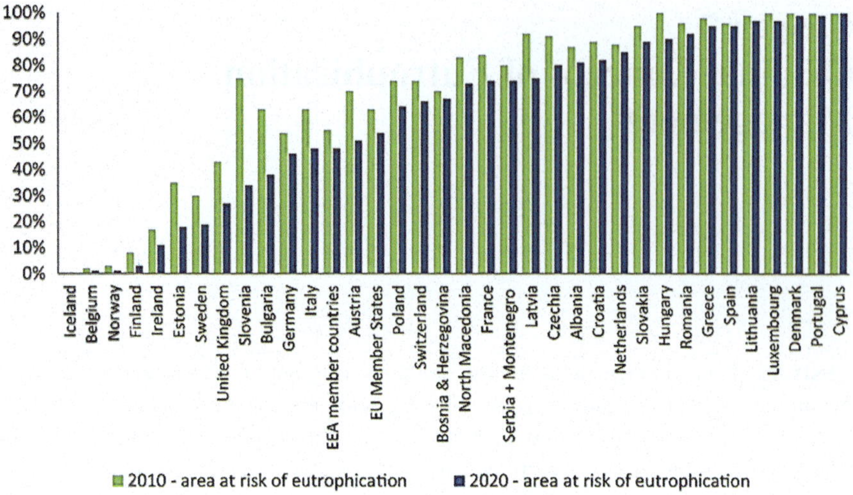

Fig. 7.1 The ecosystem area at risk of eutrophication ([2] CC BY)

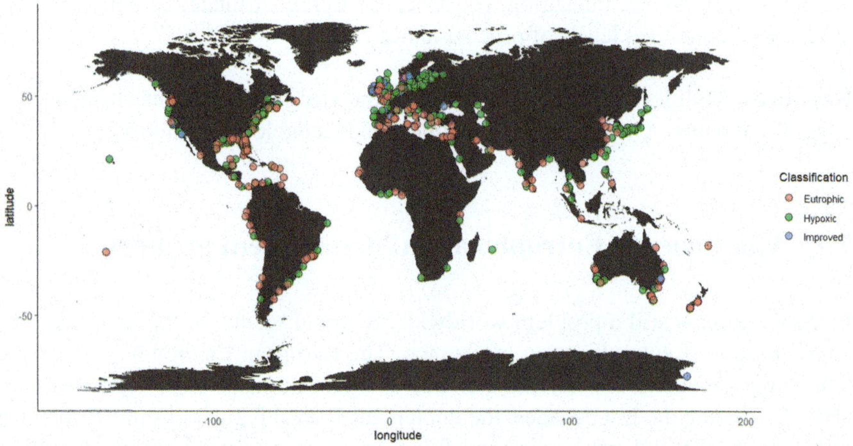

Fig. 7.2 Global eutrophication map (after Jessen [3] WRI)

has been on the role of nutrients and efforts to reduce their discharge into water bodies as a tool for influencing the ecosystem state. This remains important, but does not cover all possible approaches to the problem [4]. This is a fairly traditional method of influence, which is well studied and widely used Currently, the measures to combat eutrophication are based mainly on limiting the supply of biogenic substances to surface waters from different sources, primarily point sources.

However, eutrophication and its consequences is a large-scale biochemical process and depends on a number of internal and external factors, which are rarely taken into account in standard approaches to managing this phenomenon.

The main shortcomings include:

(1) Lack of a single definition, which leads to misunderstanding the essence of the process and confusion of causes and effects. This, in turn, leads to low effectiveness of eutrophication monitoring and management systems based on definitions, that do not reflect the essence of its mechanisms. "Our mission is to propose a better definition of eutrophication and to link the definition with monitoring and assessment systems" [5].
(2) Lack of a balance approach to understanding the process. Biotic balance seems to be the only basis, on which a system of water trophic types classification can be developed and have to be a universal criterion of the trophic status of waters and the foundation for ecological classification of water ecosystems.
(3) Eutrophication abatement strategies focus mainly on nutrient loads limiting from wastewater treatment plants, predominantly neglecting other significant sources. Wastewater treatment process becoming more and more efficient at removing nutrients, which leads to the use of very expensive, energy-intensive technologies and an increase in the carbon footprint of wastewater treatment plants, but the target effect could not be achieved.
(4) Lack of consideration of the role of bioavailable forms of biogenic substances. Consideration of this knowledge constitutes the basis for the selection of appropriate wastewater treatment technologies with low eutrophication potential, ensuring a reduction of their impact on the development of eutrophication, optimization of energy consumption and the costs of wastewater treatment.
(5) Lack of regional standardization of permissible loads in accordance with the assimilative capacity of water bodies. Water ecosystems are capable for self-sustaining and self-regulation based on their natural properties and balance and resistant to changing impacts. This ability is supported by regulatory or control mechanisms of self-purification, the set and significance of which may vary in water bodies of various types and in various climatic conditions. Permissible nutrient loads should be determined depending on the assimilative capacity of water ecosystems with consideration of total load from all the sources, setting reasonable charge amounts for each source .

(6) Ignoring the phenomenon of so called "synergistic summation", which means that is equally important to act in two directions simultaneously. Firstly, it is important to reduce the loads of nutrients. Secondly, it is equally important to reduce the input of those pollutants that weaken the ecosystems self-purification mechanisms and the potential of the ecosystem to remove phytoplankton biomass by filter feeders, e.g. surfactants and detergents This concept can be called a synecological solution [6]. A necessary element of eutrophication abatement strategy should be the preservation of the self-purification potential of reservoirs and watercourses and control pollutants loads within the assimilative capacity of ecosystems and their ecological reserve.
(7) Lack of uniform eutrophication assessment methodology, problems with monitoring of eutrophication. When developing a monitoring system, it is necessary to strive for its optimality. On the one hand, the observation system should not provide redundant information, on the other hand, the information must be representative. The degree of monitoring informativeness is directly related to all components of the observing system. The wider the list and frequency of monitored parameters, the higher the degree of information about the state of the ecosystem, however, this relationship is nonlinear. The RWD monitoring methodology, assuming the assessment of ecological/trophic state once every 4 years in descriptive categories of ecological conditions (very good, good, moderate, bad, etc.) does not allow for a reliable assessment of the actual trophic state and does not provide a reliable basis for taking effective preventive organizational/technical measures and management strategies.
(8) The use of indicators not adapted to the specificity of local conditions and the functioning of aquatic ecosystems often lead to unreliable/erroneous conclusions that cannot constitute the basis for effective management strategies.
(9) Lack of application of agricultural practices, air protection, natural barriers. This is the fundamental difference between urban and agricultural sources of nutrient loads. The first one is a typically anthropogenic nutrients source and requires drastic measures to prevent the discharge of municipal and industrial wastewaters. The industrial livestock sector has similar features due to the increasing concentration of livestock and the use of intensive technologies. The agricultural nutrients sources is a separately considered system, since it mainly preserves the mechanism of natural migration of nutrients. Under conditions of intensive agricultural development, the natural nutrient cycles change, the established mechanism of their cycles is disrupted, which is especially characteristic of the main factors of eutrophication—nitrogen and phosphorus. Reducing the supply of biogenic substances from agricultural land is a very complex, multi-level task, taking into account the rationalization and ecologization of agriculture, engineering, technical, organizational and legal methods, which is practically not implemented at the current stage.
(10) Legislation aspects. Taking into account the above significant shortcomings in management strategies for eutrophication it needs the revision of environmental legislation and introduction of the new elements into its interpretation—both national and international legislative and legal documents.

(11) Lack of comprehensive management strategies—management crisis.

7.2 Multilevel Management Approach

Taking as a starting point the balance approach to eutrophication, which is currently one of the most priority global problems, it should be understood as a kind of disturbance of the aquatic ecosystems ecological balance. This is why the solution to this problem should be based not on a separate ecosystem level, but on a system approach, widely used in modern ecology when learning about the functioning of multi-level natural ecosystems interconnected with anthropogenic systems. The aim of such approach is to deeply understand the functioning of aquatic ecosystems in conditions of increased anthropogenic pressure combined with destabilization of climatic conditions and to develop appropriate tactics and management strategies to optimize their functioning. In such conditions, the ecological approach is transformed into a science of searching for ways for humanity to survive.

Ecosystems, especially aquatic ecosystems, are characterized by unprecedented structural complexity, multi-level nature and interconnectedness of numerous factors. Controlling functions and feedback loops are formed within the systems and not directed externally. The laws and regularities of water ecosystem functioning are multi-factorial, complex and non-linear.

System analysis was recognized only in the second half of the twentieth century, but currently its wide application is supported by the development of instrumental and remote methods, computer and simulation techniques. The ecosystem approach is far from homogeneous. It has different directions, distinguished by problems formulation and the methods of solving them: structural direction (focus on the living components of the system) and functional direction (focus on transformation of matter and energy in ecosystems) [7].

The authors of the monograph are the adherents of a functional system approach to understanding the phenomenon of eutrophication. It makes possible to present a holistic, integrated assessment of the results of the entire aquatic ecosystem functioning based on the processes of matter and energy transformation.

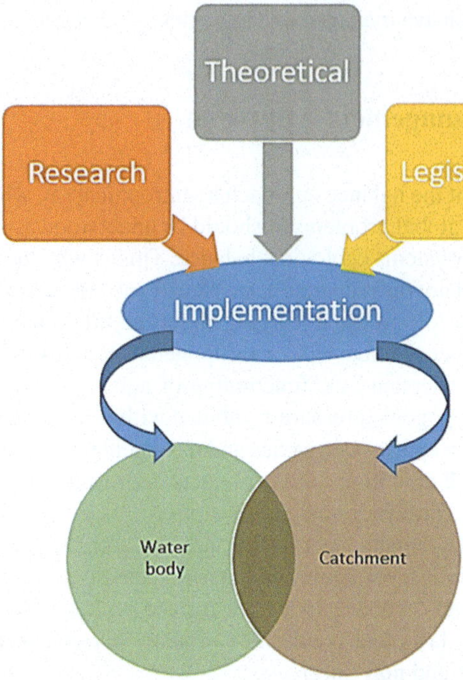

Fig. 7.3 Schematic diagram for managing the eutrophication process (the colors in the table and this figure are related)

Starting from the basic assumptions of the functional direction of the system approach, the authors developed the holistic multilevel concept of eutrophication process managing, which takes into account the aspects currently ignored in routine practices of managing this process. The concept covers coordinated activities' totality at various hierarchical levels of process management: theoretical knowledge, comprehensive research on process mechanisms, development of legal regulations based on knowledge and research results, development of organizational, technical and technological solutions in the catchment area and within water ecosystem in order to manage anthropogenic loads and maintain ecological balance. The idea of multilevel concept for eutrophication management is presented at Fig. 7.3 with detalization in Table 7.1.

Table 7.1 Concept of multilevel management strategy

Levels of decision-making	Basic activities	Undertakings
	Theoretical	1. Knowledge and understanding of the eutrophication process mechanisms based on a holistic approach to the problem 2. Determining cause—effect relationships 3. Adopting a definition reflecting the cause—effect relationship 4. Identification of stressors and responses of the aquatic ecosystem
	Research	1. Determining ecosystem-catchment relationships 2. Identification of external and internal sources of pollution (biogenic matter and their loads, other pollution reducing the self-purification capacity) 3. Evaluation of the ecological capacity of the aquatic ecosystem 4. Establishing permissible amounts of pollutant loads for each source
	Legislation	1. Regional approach to the development of ecological standards for water and wastewater quality 2. Optimization of monitoring system: integral indicators—operational monitoring—forecasting process dynamics
	Implementation (water body)	1. Prior limiting factors assessment for individual water ecosystems 2. Optimization of technical/technological methods of water protection e.g. wastewater treatment with consideration of bioavailable forms of nutrients in treated wastewater discharges (using simulation tools to select optimal treatment technologies)
	Implementation (catchment)	1. Catchment management: greening and rationalization of agricultural activities, storm water management, water protection zones etc. 2. General water protection (limiting pollutants affecting self-cleaning mechanisms and increasing sensitivity to eutrophication, water protection zones) 3. Remediation/reclamation of water ecosystems

References

1. Preisner M, Neverova-Dziopak E, Kowalewski Z (2020) An analytical review of different approaches to wastewater discharge standards with particular emphasis on nutrients. Environ Manag 66(4):694–708. https://doi.org/10.1007/s00267-020-01344-y
2. EC (2021) Economic data related to the implementation of the WFD and the FD and the financing of measures—final report. Publications Office. https://doi.org/10.2779/163850

3. Jessen C (2013) Effects of simulated eutrophication and overfishing on coral reef invertebrates, algae and microbes in the Red Sea
4. Shiklomanov I, Rodda J (2004) World water resources at the beginning of the twenty-first century, p 13
5. Andersen JH, Schlüter L, Ærtebjerg G (2006) Coastal eutrophication: recent developments in definitions and implications for monitoring strategies. J Plankton Res 28(7):621–628. https://doi.org/10.1093/plankt/fbl001
6. Ostroumov SA (2003) Anthropogenic effects on the biota: towards a new system of principles and criteria for analysis of ecological hazards. Riv Biol. 96(1):159–169. PMID: 12852181
7. Baranovskaya N, Usmanova I, Matvienko A (2013) Modern problems of ecology and environmental management. Tomsk Polytechnic University

Conclusion

According to the laws of ecology, the causes of anthropogenic environmental crises are exponential growing of human population, unlimited exploitation of natural resources and increasing environmental pollution. Nowadays, nature conservation has become the most important problem, on the correct solution of which the prosperous existence of mankind depends. Society cannot influence nature without taking into account the possible negative consequences of its economic activities. In this respect, the security of humanity and the prospects of its further existence are entirely determined by the state of the aquatic environment.

Water is critical for sustainable development, as well as for socio-economic development, energy and food production, ecosystems conservation and human survival in general. Such a unique resource as water is largely determines environmental safety and economic well-being. Considering that fresh waters resources are limited, water will become a limiting factor of economic development, with all the ensuing consequences. Sustainable Development Goal 6 (SDG 6) is to ensure the availability and sustainable management of water and sanitation for all. According to the resolution adopted by the UN General Assembly in July 2010, every person has the right to access to sufficient water for personal and domestic use. At the same time, water must be safe, acceptable in quality and price. Water is also at the heart of adaptation to climate change, providing a critical link between society and the environment.

Currently, anthropogenic eutrophication is considered as the most important factor in the negative impact of human activity on water bodies, and research of the mechanisms of this phenomenon development is of paramount importance. The process of eutrophication of surface water bodies is influenced by a number of factors: biotic, abiotic, anthropogenic. Due to the large number of interdependences and interactions between anthropogenic and natural systems, there is concern about the ever-increasing anthropogenic load, which not only enhances the effect of the anthropogenic factor in the process of surface water eutrophication, but also affects the rest of the complex of factors. Management and regulation of economic activities near

E. Neverova-Dziopak and Z. Kowalewski, *Eutrophication: A Global Environmental Problem*, Springer Water, https://doi.org/10.1007/978-3-031-83926-9

water bodies will allow influencing this process through a complex of eutrophication factors.

Obviously, such an approach to solving the problem requires breaking a number of stable thinking stereotypes, that have developed over many decades. In conditions of rapid urbanization, population growth and industrialization, the natural environment can be preserved only with the purposeful creation of special systems of engineering and environmental development that regulate the state of ecosystems. In modern conditions, human activity should be considered not only as the main cause of degradation, but also as the only factor whose impact can prevent the development of this process.

The manufacturer's authorised representative in the EU is Springer Nature Customer Service Centre GmbH, Europaplatz 3, 69115 Heidelberg, Germany. If you have any concerns regarding our products, please contact ProductSafety@springernature.com

Printed and bound by CPI Group (UK) Ltd, Croydon, CR0 4YY
30/09/2026
02225604-0004